A Digital Signal Processing Laboratory using the TMS320C30

Henrik V. Sorensen
Ariel Corporation

Jianping Chen
Nantong Textile Engineering Institute

Prentice Hall, Upper Saddle River New Jersey 07458

Library of Congress Cataloging-in-Publication Data

Sorensen, Henrik V.
A Digital Signal Processing Laboratory Using the TMS320C30 /
Henrik V. Sorensen and Jianping Chen
p. cm.
Includes bibliographical references and index.
ISBN: 0-13-741828-0
1. Digital signal processing. 2. Electric filters, Digital.
3. Texas Instruments TMS320C30 4. MATLAB. I. Chen, Jianping.
II. Title.
CIP Data Available

Acquisitions editor: **ALICE DWORKIN**
Editor-in-Chief: **MARCIA HORTON**
Managing editor: **BAYANI MENDOZA DELEON**
Director of production and manufacturing: **DAVID W. RICCARDI**
Production editor: **IRWIN ZUCKER**
Copy editor: **LEABE BERMAN**
Cover designer: **BRUCE KENSELAAR**
Buyer: **DONNA SULLIVAN**

Simon & Schuster / A Viacom Company
Upper Saddle River, New Jersey 07458

Printed in the United States of America

10 9 8 7 6 5 4 3 2

ISBN 0-13-741828-0

Prentice-Hall International (UK) Limited, London
Prentice-Hall of Australia Pty. Limited, Sydney
Prentice-Hall Canada Inc., Toronto
Prentice-Hall Hispanoamericana, S.A., Mexico
Prentice-Hall of India Private Limited, New Delhi
Prentice-Hall of Japan, Inc., Tokyo
Simon & Schuster Asia Pte. Ltd., Singapore
Editora Prentice-Hall do Brasil, Ltda., Rio de Janeiro

Contents

Preface

This book was developed over a number of years from the notes written for a DSP Laboratory course that was taught at the University of Pennsylvania, Philadelphia. Centered around a set of experiments for the TMS320C30, the goal of the book is to teach how to program the TMS320C30 and illustrate important concepts from the theory of digital signal processing. To efficiently use a DSP processor, the user must have a solid understanding of DSP algorithms as well as an appriciation of basic computer architecture concepts. We will assume the reader has this knowledge and will seldom derive any equations. The material is targeted at first-year graduate students, as well as seniors, with a good background in digital signal processing. Many good books have been written in this area, but we recommend *Discrete-time Signal Processing*[9], since we use the same notation. For the more advanced material not covered there, we relied on *Handbook for Digital Signal Processing*[8].

The book is developed for a one semester graduate course with a suggested outline:

- one week devoted to the TMS320C30, its instruction set, and the EVM board
- six weeks devoted to Chapters 3–8, with one chapter covered per week
- one week devoted to either Chapter 9 or 10
- five weeks devoted to two-person projects as shown in Chapter 11.

To help the readers along, many chapters contain sample code that forms the foundation for the exercises. Some of these exercises involve minor modifications of the example code, while others are more elaborate. By studying and executing the example code, the book can also successfully be used as a self-study guide or for setup and evaluation for commercial applications.

During the years when these notes was used, the students who took the course provided invaluable feedback. The authors would like to acknowledge the students of this course, and John Schierling in particular, for providing many of the corrections that eventually turned the notes into a "real" book. We would also like to acknowledge Daniel Charnitsky, Ximing Chen, Nageen Himayat, Anestis Karasaridis, Michael Meixler, Joe Palovick, James Tau, Dean Thompson, and Philip Trahanas for providing the projects presented in Chapter 11. The crew at Prentice Hall (Alice Dworkin, Tom Robbins, and Irwin Zucker) deserve praise for their helpful assistance (and "Trotter excursion!") during this whole project. Finally the authors would like to acknowledge the support from Texas Instruments and National Science Foundation under grant NCR 90-16165.

Chapter 1

Introduction

Processors for Digital Signal Processing (DSP) have changed dramatically since 1983, when Texas Instruments announced the TMS32010—the first commercially available DSP processor. By today's standards, the TMS32010 is rather slow with a 200 ns cycle time and has a very limited data memory, with only 288 bytes. It uses a fixed-point data representation that can cause overflow unless scaling is performed.

The TMS320C30 was introduced in 1988 as the first floating-point device in the TMS320 family and is more that three times faster than its predecessor, with a 60 ns cycle time. It has 8 kbytes of internal memory and supports floating-point arithmetic that can achieve a peak performance of 33 Mflops. It also has an extensive set of internal peripherals, such as DMA, timers, serial ports, etc., that makes it ideal for digital signal processing.

To create efficient, real-time programs for a DSP processor, like the TMS320C30, requires an understanding of the processor instruction set, hardware configuration, software development tools, and hardware development tools. All this information is contained in the manufacturer's manuals [19, 17, 15, 18], but they are somewhat intense and not structured for teaching purposes. We will hence give a brief introduction of the TMS320C30 processor, the evaluation module (EVM), and the TMS320C30 software tools. We will focus on the essential features necessary for the first few experiments, but the description is not always complete and the user may have to supplement with information from the manuals. There are many related books dealing with earlier DSP chips from the Texas Instruments TMS320 family, such as [6, 5, 2].

In Chapter 2, we briefly introduce the features of the TMS320C30, the instruction set, the assembly language tools, the EVM board and the C

Directory	Description
C30_LAB	All TMS320C30 files listed in this book
C30_LAB\SOL	Solutions to selected problems
C30_LAB\DATA	Filter coefficients for the FIR and IIR section
C30_LAB\MATLAB	Matlab support files used through-out the book
C30_LAB\PROJ	Source code for projects in Chapter 11

Table 1.1: Content of included disk

source debugger. It puts emphasis on a few important points, rather than trying to give a complete overview. Chapter 3 presents a simple sampling program, which samples an input signal via the A/D port and plays it back via the D/A port. This program will form the foundation for the experiments in the following seven chapters. Chapter 4 presents three different waveform geenerators: squarewave, sinewave, and random number wave. Chapter 5 and 6 deals with implementation of FIR and IIR filters, respectively. Chapter 7 focuses on Fourier transforms and real-time spectrum analysis. Chapter 8 talks about quantization effects of both floating-point and fixed-point systems. Chapter 9 and 10 focuses on implementation of adaptive and multi-rate systems. The final chapter of the book gives a series of project suggestions and describes five example projects in detail.

In the back of this book you will find a disk that contains all programs listed in this book. It also contains solutions to some of the problems, other support files, as well as source code for the projects listed in Chapter 11. Installation instructions can be found in the text file READ.ME, which is included on the disk.

Chapter 2

The TMS320C30 Digital Signal Processor

The TMS320 series is a family of single-chip processors optimized for digital signal processing and other high-speed numeric processing applications. The processors are fully programmable, just as general micro processors, and they achieve their speed by implementing functions in hardware that regular microprocessors do through software or micro code.

2.1 Features of the TMS320C30

Introduced in 1988, the TMS320C30 is the third generation of the family. It has many architectural features that permit very efficient implementations of DSP algorithms. Figure 2.1 shows the TMS320C30 architecture.

Some key features of the device are listed below.

- 60 ns single-cycle instruction execution time that can achieve 33.3 MFLOPS (million floating-point operations per second) and 16.7 MIPS (million instructions per second)
- One 4k * 32-bit single-cycle dual-access on-chip ROM block
- Two 1k * 32-bit single-cycle dual-access on-chip RAM block
- 64 * 32-bit instruction cache
- 32-bit instruction and data words, 24 bit addresses
- 40/32-bit floating-point/integer multiplier and ALU
- 32-bit barrel shifter

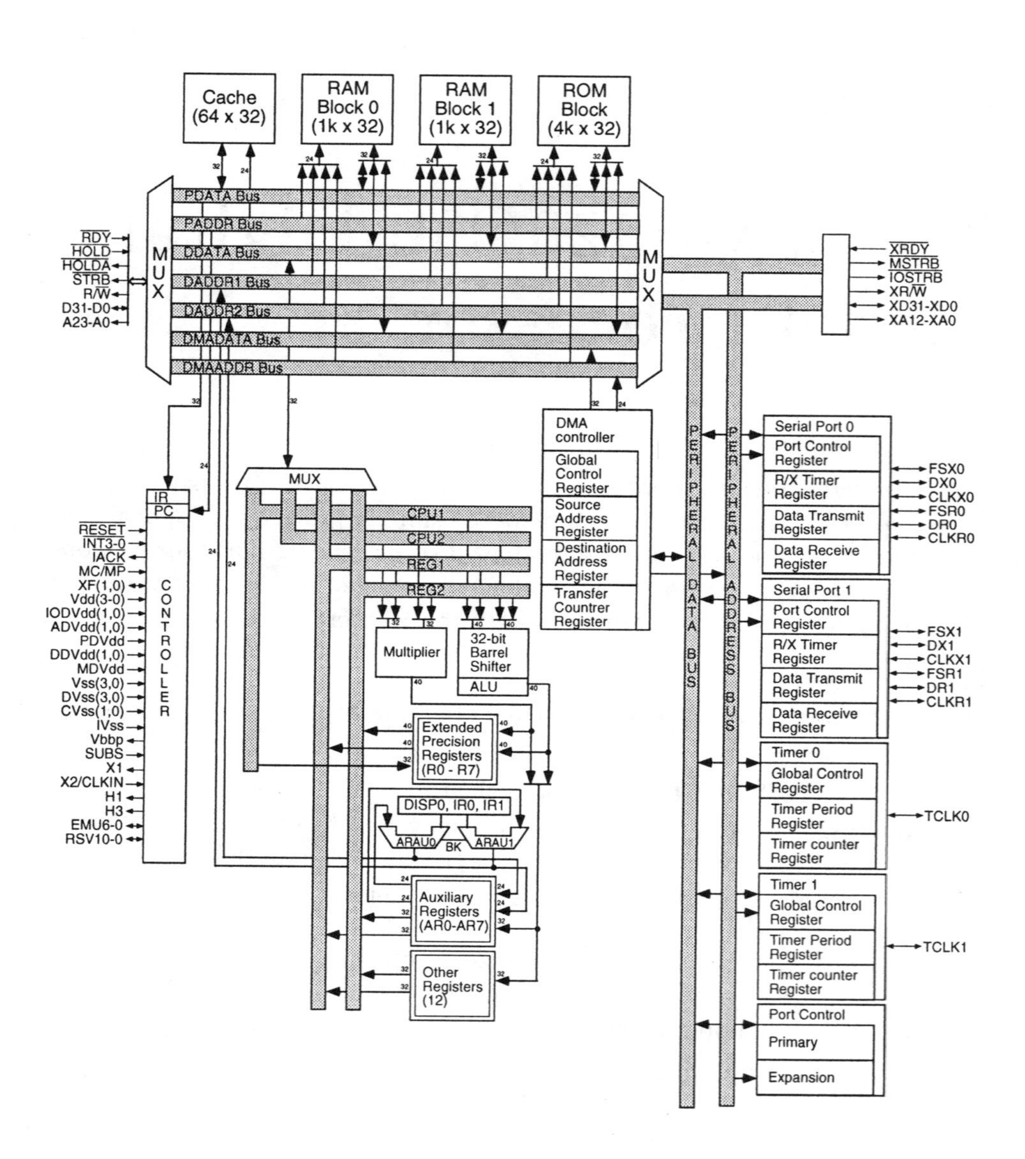

Figure 2.1: TMS320C30 Block Diagram

- Eight extended-precision register (accumulators)
- Two address generators with eight auxiliary registers and two auxiliary register arithmetic units
- On-chip Direct Address Access (DMA) controller for concurrent I/O and CPU operation
- Integer, floating-point and logical operations
- Two- and three-operand instructions
- Parallel ALU and multiplier instructions in a single cycle
- Block repeat capability
- Zero-overhead loops with single-cycle branches
- Conditional calls and returns
- Interlocked instructions for multiprocessing support
- Two serial ports to support 8/16/32-bit transfer
- Two 32-bit timers
- Two general-purpose external flags, four external interrupts
- 180-pin gird array (PGA) package; 1 μm CMOS

The TMS320C30 is similar to regular microprocessors in many ways, but it also has some unique features that make it highly suitable for digital signal processing purposes. One of these features is its Harvard architecture, which means that the chip has separate buses for program and data. Thus, instructions can be fetched, even while data is being accessed. Another feature of great importance is the hardware multiplier, which executes in a single cycle and hence makes a multiplication and an addition equally fast.

The two most salient characteristics of the TMS320C30 are its high speed (60 nanoseconds cycle-time) and floating-point arithmetic capability. The high speed makes the implementation of real-time applications easy, even when the other architecture advantages are not considered. For example, at the 8 kHz sampling rate, commonly used in voice-quality applications, over two thousand single-cycle instructions can be achieved between two samples. Each instruction, by itself, requires four cycles from fetch to execution, but a very efficient pipeline can execute them in a single cycle under very mild restrictions.

The floating-point capability permits the handling of numbers with a very high dynamic range without worrying about overflows. In many basic DSP routines, such as a FFT, the computed values tend to increase from one stage to the next; whence fixed-point arithmetic will most definitely cause overflow if the incoming numbers are "full-scale" and no provisions are made for scaling. All these considerations are eliminated with the floating-point capability of the TMS320C30. Quantization effects also become less of a problem with floating-point arithmetic. Floating-point instructions on the TMS320C30 are performed with the same speed as any other operation so there is no direct performance degradation associated with floating-point operations. (Given a constant chip area and technology, however, one can normally get more MIPS out of a fixed-point device than a floating point device.)

The TMS320C30 has eight extended-precision (40 bit) registers, `R0` – `R7`, that can be used as accumulators, and eight auxiliary (32 bit) registers, `AR0` – `AR7`, that can be used for addressing and integer arithmetic. In many applications, these registers provide sufficient temporary storage of internal variables that there is no need for external (memory) storage. Performing arithmetic between these registers (as opposed to, between memory locations) normally increases the program efficiency significantly. The two index registers, `IR0` and `IR1`, are used for indexing/incrementing the contents of the auxiliary registers, `AR0` – `AR7`, and make the indirect addressing mode very flexible.

A very powerful structure in the TMS320C30 is the block-repeat capability, which allows an instruction section (a block) to be repeated any number of times with no overhead for the looping control. This is accomplished by controlling and decrementing the loop-counter register, RC, from hardware rather than software. The repeated code thus behaves as if it was straight-line coded but with a program size equal to the one in the looped code. In this way, the most often performed loop of an algorithm, like the FFT butterfly, can be implemented in a block-repeat form, which saves execution time while preserving the clarity of the program and limiting the program size.

The TMS320C30 supports a variety of different addressing modes to allow a flexible access of data from memory, registers, and the I/O ports. It has six different address types that can be applied to each of the five different addressing modes. Furthermore, two special indirect addressing modes, circular and bit-reversed addressing, are provided to meet the need of basic DSP routines. Many algorithms, such as convolution and correlation, require the implementation of a circular buffer in memory. The circular buffer is used to implement a sliding window that contains the most recent data to be processed. As a new data element is brought in, it overwrites the oldest data element. The circular addressing mode takes care of all this with a minimal setup. The bit-reversed addressing mode eliminates the need for swapping data elements at the beginning or at the end of FFT algorithms. Using this addressing mode, a sequence of data points is accessed in bit-reversed, rather than sequential, order, and the reordering of the data can be done during the retrieval of the data instead of through a piece of dedicated code, thus saving both time and space.

Another speed boost is achieved through the device's parallel capability. The TMS320C30 can perform parallel multiply and ALU operations on integer or floating-point data in a single cycle. Although there are some restrictions on the addressing modes, the parallel instruction set includes multiply/add, load/load, load/store, or add/store, which are very effective for implementing DSP algorithms. The multiply/add instruction, for example, can be used to implement, at one time, the coefficient multiplication and the product accumulation of a filtering or FFT algorithm.

General-purpose applications are greatly enhanced by the large dual-access memories, multiprocessor interface, internally and externally generated wait states, two timers, two serial ports, one DMA channel supporting concurrent I/O, and multiple interrupt structure. The TMS320C30 supports a wide variety of system applications from host processor to dedicated coprocessor.

Further details can be found in the "TMS320C3x User's Guide" [19].

2.2 The TMS320C30 Instruction Set

2.2.1 The Instruction Set

The TMS320C30 assembly language instruction set supports numerical intensive signal processing and general-purpose instructions. All instructions consist of a single machine word of 32 bits, and most instructions take one single cycle to execute. The instruction set contains 114 instructions organized into the following function groups:

- Load and store instructions
- Arithmetic instructions
- Logical instructions
- Program control instructions
- Parallel instructions
- Interlocked instructions

Load and Store Instructions

The load and store instruction group consists of 12 instructions, listed in Table 2.1. These instructions can load a word from memory into a register,

Instruction	Description	Instruction	Description
LDE	Load floating-point exponent	POP	Pop integer from stack
LDF	Load floating-point value	POPF	Pop floating point value from stack
LDF*cond*	Load floating-point value conditionally	PUSH	Push integer on stack
LDI	Load integer	PUSHF	Push floating-point value on stack
LDI*cond*	Load integer conditionally	STF	Store floating-point value
LDM	Load floating-point mantissa	STI	Store integer

Table 2.1: Load and Store Instructions

store a word from a register into memory, or manipulate data on the system stack. NOTE: When pushing (or poping) a 40 bit accumulator, the high and low part of the register must be pushed (or poped) separately thus requiring two instructions.

Arithmetic Instructions

The TMS320C30 supports a complete set of arithmetic instructions. These instructions provide integer operations, floating-point operations, and multiprecision arithmetic. While most instructions have only two operands, the TMS320C30 also supports several three-operand arithmetic and logical operations. This allows the TMS320C30, in a single cycle, to read two operands from memory or the register file, as well as store the result. Table 2.2 lists the arithmetic instructions. The instructions that have two- and three-operand versions are marked with "†."

Logical Instructions

The TMS320C30 supports a complete set of logical instructions, some of which have three-operand versions. They are summarized in Table 2.3.

Program Control Instructions

The program control instruction group consists of all of those instructions that affect program flow. The repeat modes allow a non-overhead repetition of a block or single line of code. Both standard and delayed (single-cycle) branching are supported. Table 2.4 lists the program control instructions.

Instruction	**Description**	**Instruction**	**Description**
ABSF	Absolute value of a floating-point value	NEGB	Negate integer with borrow
ABSI	Absolute value of an integer	NEGF	Negate floating-point value
ADDC †	Add integer with carry	NEGI	Negate integer
ADDF †	Add floating-point values	NORM	Normalize floating-point value
ADDI †	Add integers	RND	Round floating-point value
ASH †	Arithmetic shift	SUBB †	Subtract integer with borrow
CMPF †	Compare floating-point values	SUBC	Subtract integer conditionally
CMPI †	Compare integers	SUBF †	Subtract floating-point values
FIX	Convert floating-point to integer	SUBRB	Subtract reverse-integer with borrow
FLOAT	Convert integer to floating-point	SUBRF	Subtract reverse floating-point value
MPYF †	Multiply floating-point values	SUBRI	Subtract reverse integer
MPYI †	Multiply integers		

† Two and three operand versions

Table 2.2: Arithmetic Instructions

Instruction	Description	Instruction	Description
AND †	Bitwise logical AND	ROLC	Rotate left through carry
ANDN †	Bitwise logical AND with compliment	ROR	Rotate right
LSH †	Logical shift	RORC	Rotate right through carry
NOT	Bitwise logical compliment	TSTB †	Test bit fields
OR †	Bitwise logical OR	XOR †	Bitwise exclusive OR
ROL	Rotate left		

† Two and three operand versions

Table 2.3: Logical Instructions

NOTE: You can not branch (using Bxx or BR[D]) within a repeat block loop thus only the inner-most-loop can use repeat block looping.

Parallel Instructions

The TMS320C30 supports a group of parallel instructions that allow a high degree of parallelism. Some of the TMS320C30 instructions can occur in pairs that will be executed in parallel. These parallel instructions provide parallel loading of registers, parallel arithmetic operations, and arithmetic or logical instructions used in parallel with a store instruction. Table 2.5 lists the valid instruction pairs. NOTE: When the same register is used both as a source and destination within a parallel instruction, the register-value is not updated until the end of the entire instruction, thus all sources use the "old" value.

Interlocked-Operation Instructions

The interlocked-operation instructions support multiprocessor communication in a shared memory bank environment. Table 2.6 lists the five interlocked-operation instructions.

Conditional Codes and Flags

The TMS320C30 supports conditional loads, branches, trap calls, and returns. Twenty condition codes are provided that can be used with any

Instruction	**Description**	**Instruction**	**Description**
B*cond*[D]	Branch conditionally standard or delayed	NOP	No operation
BR[D]	Branch unconditionally standard or delayed	RETI*cond*	Return from interrupt conditionally
CALL	Call subroutine	RETS*cond*	Return from subroutine conditionally
CALL*cond*	Call subroutine conditionally	RPTB	Repeat block of instructions
DB*cond*[D]	Decrement and branch conditionally	RPTS	Repeat single instruction
IDLE	Idle until interrupt	TRAP*cond*	Trap conditionally
SWI	Software interrupt		

Table 2.4: Program Control Instructions

of the conditional instructions, such as RETS*cond* or LDF*cond*. The conditions include signed and unsigned integer comparisons, comparisons to zero, and comparisons based on the status of individual condition flags. The condition codes and flags used are listed in Table 2.7.

2.2.2 The Addressing Modes

The TMS320C30 supports five groups of different addressing modes :

- General addressing mode
- Three-operand addressing mode
- Parallel addressing mode
- Long-immediate addressing mode
- Conditional-branch addressing mode

Within each group, six types of addressing may be used, which allow access of data from memory, registers, and the instruction word:

- Register
- Direct
- Indirect
- Short-immediate
- Long-immediate

Instruction	**Description**	**Instruction**	**Description**
	Parallel Arithmetic with Store Instructions		
ABSF ||STF	Absolute value of a floating-point	MPYF3 ||STF	Multiply floating-point
ABSI ||STI	Absolute value of an integer	MPYI3 ||STI	Multiply integer
ADDF3 ||STF	Add floating-point	NEGF ||STF	Negate floating-point
ADDI3 ||STI	Add integers	NEGI ||STI	Negate integer
AND3 ||STI	Bitwise logical-AND	NOT3 ||STI	Bitwise compliment
ASH3 ||STI	Arithmetic shift	OR3 ||STI	Bitwise logical-OR
FIX ||STI	Convert floating-point to integer	STF ||STF	Store floating-point
FLOAT ||STF	Convert integer to floating-point	STI ||STI	Store integer
LDF ||STF	Load floating-point	SUBF3 ||STF	Subtract floating-point
LDI ||STI	Load integer	SUBI3 ||STI	Subtract integer
LSH3 ||STI	Logical shift	XOR3 ||STI	Bitwise exclusive-OR
	Parallel Load Instructions		
LDF ||LDF	Load floating-point	LDI ||LDI	Load integer
	Parallel Multiply and Add/Subtract Instructions		
MPYF3 ||ADDF3	Multiply and add floating-point	MPYI3 ||ADDI3	Multiply and add integer
MPYF3 ||SUBF3	Multiply and subtract floating-point	MPYI3 ||SUBI3	Multiply and subtract integer

Table 2.5: Parallel Instructions

Instruction	Description	Instruction	Description
LDFI	Load floating-point value, interlocked	STFI	Store floating-point value, interlocked
LDII	Load integer, interlocked	STII	Store integer, interlocked
SIGI	Signal, interlocked		

Table 2.6: Interlocked-Operation Instructions

- PC-relative

Below is a brief description of the different addressing modes. A more detailed explanation can be found in Section 6 of the "Third Generation TMS320 User's Guide" [19].

General Addressing Modes

Instructions that use general addressing modes are general-purpose instructions, such as ADDI, MPYF, and LSH. Four types of addressing may be used for these instructions.

Register mode: The operand is a CPU register. For floating-point operations, use an extended register (R0 - R7). For integer operations, use any register.

Short-immediate mode: The operand is 16-bit immediate value, which may be signed integers, unsigned integers, or floating-point numbers, depending on the instruction.

Direct mode: The operand is the contents of a 24-bit address, specified by @addr. The 8 MSBs of the address are specified by the DP register; the 16 LSBs are specified by the instruction word.

Indirect mode: The address of the operand is indicated by an auxiliary register. Table 2.8 lists the various forms that indirect operands may take. The displacement may be specified as a value from 0-255 or as one of the index registers (IR0 or IR1). The displacement of one can be a default.

Three-Operand Addressing Modes

Three-operand addressing modes are used with the three-operand instructions, such as LSH3, ADDF3, or CMPI3. Two types of addressing can be used as the three-operand addressing modes.

Condition	Code	Description	Flags
Unconditional Compare			
U	00000	Unconditional	don't care
Unsigned Compare			
LO	00001	Lower than	C
LS	00010	Lower or same	C OR Z
HI	00011	Higher than	$\overline{C}$ AND $\overline{Z}$
HS	00100	Higher or same	$\overline{C}$
EQ	00101	Equal	Z
NE	00110	Not equal	$\overline{Z}$
Signed Compare			
LT	00111	Less than	N
LE	01000	Less than or equal	N OR Z
GT	01001	Great than	$\overline{N}$ AND $\overline{Z}$
GE	01010	Great than or equal	$\overline{N}$
EQ	00101	Equal	Z
NE	00110	Not equal	$\overline{Z}$
to Zero			
Z	00101	Zero	Z
NZ	00110	Not zero	$\overline{Z}$
P	01001	Positive	$\overline{N}$ AND $\overline{Z}$
N	00111	Negative	N
NN	01011	Non-negative	$\overline{N}$
Compare to Condition Flags			
NN	01011	Non-negative	$\overline{N}$
N	00111	Negative	N
NZ	00110	Non-zero	$\overline{Z}$
Z	00101	Zero	Z
NV	01100	No overflow	$\overline{V}$
V	01101	Overflow	V
NUF	01110	No underflow	$\overline{UF}$
UF	01111	Underflow	UF
NC	00100	No carry	$\overline{C}$
C	00001	Carry	C
NLV	10000	No latched overflow	$\overline{LV}$
LV	10001	Latched overflow	LVF
NLUF	10010	No latched floating-point underflow	$\overline{LUF}$
LUF	10011	Latched floating-point underflow	LUF
ZUF	10100	Zero or floating-point underflow	Z OR UF

Table 2.7: Conditional codes and flags

Operand	Description
*ARn	Indirect with no displacement
*+ARn(disp)	Indirect with predisplacement or preindex add
*−ARn(disp)	Indirect with predisplacement or preindex subtract
*++ARn(disp)	Indirect with predisplacement or preindex add and modification
*−−ARn(disp)	Indirect with predisplacement or preindex subtract and modification
*ARn++(disp)[%] †	Indirect with postdisplacement or postindex add and modification
*ARn−−(disp)[%] †	Indirect with postdisplacement or postindex subtract and modification
*ARn++(IR0)B	Indirect with postindex (IR0) and bit-reversed modification

† Optional circular modification (specified by %)

Table 2.8: Indirect addressing Modes

Register mode: Same as for general addressing modes.

Indirect mode: Same as for general addressing modes, but the displacement is limited to 0, 1, IR0, or IR1.

Parallel Addressing Modes

Instructions that use parallel addressing (indicated by ||), such as MPY || ADD, allow for the greatest amount of parallelism possible. Like the three-operand addressing modes, register and indirect modes are used for parallel addressing.

Register mode: The operand is an extended register (R0 - R7). In some cases, only R0/R1 or R2/R3 can be used as an operand.

Indirect mode: Same as for general addressing modes, except the displacement is limited to 0, 1, IR0, or IR1.

Long-Immediate Addressing Mode

The long-immediate addressing mode is used to encode the program control instructions (BR, BRD, and CALL), for which it is useful to have a 24-bit absolute address contained in the instruction word. The operand is a 24-bit immediate value, which is usually specified by a label.

Conditional-Branch Addressing Modes

Instructions that use the conditional-branch addressing modes (such as B*cond*, CALL*cond*, DB*cond*, etc.) can perform a variety of conditional operations. Two types of addressing modes can be used for the conditional-branch addressing.

Register mode: Same as for general addressing modes; the contents of the register are loaded into the program counter.

PC-relative mode: A signed 16-bit displacement is added to the program counter. The destination address is usually specified as a label; the assembler calculates the displacement.

2.3 The Software Development Tools

Like all computers, the DSP processor gets its instructions in the form of binary numbers. The mnemonic assembly language instruction code must first be assembled into the numerical machine instruction code and then linked to form an executable program. These assembler and linker programs are available from the manufacturer. A complete set of the assembly language tools for the TMS320C30 include the assembler, linker, archiver, and object format converter. In this section, we will briefly describe the major aspects of the assembler and the linker, since they are needed throughout this book. For a full description of the tools, the manufacturer's manual [17] should be consulted.

2.3.1 The Assembler

The assembler translates assembly language source files into machine language object files. Source files can contain instructions, assembler directives, and macro directives. You can use assembler directives to control various aspects of the assembly process, such as the source listing format, data alignment, and section content.

Source Statement Format

TMS320C30 assembly language source programs consists of source statements that are written in lines. Each source statement line consists of a "four field" format. The general syntax for the source statement format is:

```
label     mnemonic        operand(s)      comment
```

Statements must begin with a label, a blank, an asterisk, or a semicolon. One or more blanks must separate each field.

Label field: A label can contain up to 32 alphanumeric characters. It serves to associate a symbolic address with a location in the program. Lines that are labeled in the assembler can then be referenced by name. This is useful for modular programming and for branch instructions. Labels are optional, but, if used, they must begin in Column 1.

Mnemonic field: The mnemonic field can contain an instruction, an assembler directive, a macro directive, or an invocation. Notice that the mnemonic field can not start in Column 1, otherwise it would be interpreted as a label.

Operand field: The operand field is a list of operands. An operand can be a constant, a symbol, or a combination of constants and symbols in an expression. Constants can be binary, decimal, hexadecimal integers, floating-point numbers, and assembly-time or character constants. Symbols include labels, register names, and the symbols defined in the assembly with assembler directives.

Comment field: Comments are optional. They are notes about the program that is significant to the programmer, but do not affect the program assembly. When starting in Column 1, a comment can begin with an asterisk or a semicolon ("*" or ";"). Comments that begin in any other column must begin with a semicolon.

The next several lines show examples of source statements:

```
SYM     .SET    0A5h            ;Set symbol SYM to 0A5 hex
START   ADDI    SYM+5, R1       ;Add (SYM+5) to contents of R1
        BR      START           ;Branch to the line labeled START
```

Assembler Directives

Assembler directives can be used to control the assembly process and to enter data (constants or initial values) into the program. They do not produce executable statements (opcodes). The TMS320C30 assembler directives are divided into several groups according to functions, such as section directives, memory initialization directives, etc. The TMS320C30 Assembly Language Tools User's Guide [17] discusses the assembler directives in detail. Some of the most important assembler directives are described below.

Section Directives

Section directives associate the various portions of an assembly language program, such as code and data, with appropriate sections.

.BSS directive: The `.bss` directive reserves space in the `.bss` uninitialized section for variables. It is usually used to allocate data into RAM for run-time variables such as I/O buffers. The `.bss` directive has the form:

```
        .bss     symbol, size_in_words
```

The symbol points to the first location of the reserved memory space. The size specifies the number of the words to be reserved in the `.bss` section. It must be an absolute expression.

.USECT directive: The `.usect` directive reserves space for variables in a named uninitialized section. This directive is similar to the `.bss` directive, but it allows you to define additional sections separately from the `.bss` section. The `.usect` directive has the form:

```
symbol    .usect     "section_name", size_in_words
```

The parameter `section_name` names the uninitialized section. It must be enclosed in double quotes. The symbol and the size have the same meaning as in the `.bss` directive.

.DATA directive: The `.data` directive tells the assembler to begin assembling source code into the `.data` section, which usually contains tables of data or preinitialized variables, such as trigonometric tables, π, or filter taps. The `.data` section is a default section. The `.data` directive has the form:

```
        .data
```

.TEXT directive: The `.text` directive tells the assembler to begin assembling source code into the `.text` section, which normally contains executable code. The `.text` section is a default section. The form of `.text` directive is:

```
        .text
```

.SECT directive: The `.sect` directive defines an initialized named section and tells the assembler to begin assembling source code into the named section. Named sections can be used for data or code that must be allocated into memory separately from the `.data` or `.text` sections. It is often used to separate large programs into logical partitions; separating, for example, the common subroutines from the main program or separating constants belonging to different tasks. The `.sect` directive has the form:

```
.sect     "section name"
```

Memory Initialization Directives

Memory initialization directives assemble data values into memory locations in the current section.

.WORD directive: The `.word` directive places one or more 32-bit integer values into consecutive words in the current section. It has the form:

```
.word     value_1 [, ..., value_n]
```

The values can be either absolute or relocatable expressions. This allows you to initialize memory with pointers to variables or labels.

.FLOAT directive: The `.float` directive places one or more 32-bit, single-precision, floating-point constants into consecutive words of the current section. It has the form:

```
.float     value_1 [, ..., value_n]
```

The value must be a floating-point constant, or a symbol that has been equated to a floating-point constant.

.SET directive: The `.set` directive assigns a value to a symbol. This type of symbol is known as an assembly-time constant; it can then be used in source statements in the same manner as a numeric constant. The `.set` directive has the form:

```
symbol    .set  value
```

The symbol must appear in the label field. The value can be an expression containing constants and symbols, but the symbols must be previously defined in the current source module.

.SPACE directive: The `.space` directive reserves a specific number of words in the current section and fills them with zeros. It has the form:

```
.space     size_in_words
```

Other Directives

Besides the sections and the memory initialization directives, the assembler has some other kinds of directives. The following two are the most often used among them.

.GLOBAL directive: The `.global` directive declares a symbol to be external so that it is available to other modules at link time. It has the form:

```
.global   symbol_1 [,..., symbols_n]
```

A symbol must be declared global when it is not defined in the current module, or it is defined in the current module but will be used in external modules. This command is often used to create symbolic labels for the C Source Debugger, which helps maneuver within big programs.

.END directive: The `.end` directive is an optional directive that terminates the assembler. It should be the last statement of a program. The assembler will ignore any source statements that follow an `.end` directive. It has the form:

```
.end
```

Invoking the Assembler

The assembler is a set of programs installed and executed on the host system. To invoke the assembler, enter the command:

```
ASM30 [input_file] [object_file] [-options]
```

The `input_file` is the name of the source code program. If you do not supply an extension, the assembler assumes that the input file has the default extension `.asm`.

The `object_file` is the name of the object file that the assembler creates. The assembler assumes a default extension `.obj` for the object file. If you do not supply an object filename, the assembler creates an object file using the input filename with the `.obj` extension.

The `options` identify the assembler options that you want to use. The most common ones are:

- The `-l` option tells the assembler to create a listing file showing where program and variables are allocated within each section.
- The `-s` option puts all symbols (not just the global ones) in the symbol table so as the debugger may access them.

- The `-c` option makes the case *insignificant* in symbol names. If not, defined case is significant.

The assembly language tools user's manual [17] describes all the options in detail.

By naming your source file with the `.asm` extension (as `filename.asm`) and having the object file with the same name, you can invoke the assembler simply by entering:

```
ASM30 filename
```

2.3.2 The Linker

The linker combines object files into a single executable program. As it creates the executable module, it performs relocation and resolves external references. The linker supports a C-like command language to control the memory configuration, the output section definition, and the address binding.

Linker Directives

The linker command language supports expression assignment and evaluation, and provides two powerful directives, MEMORY and SECTION directives. Using these directives, you can define a memory model that conforms to your target system memory, combine object file sections, allocate sections into specific memory areas, and define or redefine global symbols at link time.

MEMORY Directive

The linker determines where output sections should be allocated into memory, so it must have a model of the target memory to accomplish this task. The MEMORY directive allows you to specify a specific memory model matching your target system. You can define the types of memory your system contains and the address ranges they occupy. After you use the MEMORY directive to define a memory model, you can use the SECTIONS directive to allocate the assembler output sections into the memory sections.

The MEMORY directive identifies ranges of memory that are physically present in the target system and can be used by a program. Each memory range has a name, a starting address, and a length. The general form of the MEMORY directive is:

```
MEMORY
{
   name_1 [(attr)]:     origin=constant, length=constant
      .                                    .
      .                                    .
      .                                    .
   name_n [(attr)]:     origin=constant, length=constant
}
```

The label `name_1` names a memory range, such as ROM or RAM. The `attr` parameter specifies optional attributes associated with the named range. The `origin` specifies the starting address of a memory range. The `length` specifies the number of words in the named memory range. The `constant` may be a decimal or hexadecimal integer.

The MEMORY directive is used in a linker command file. Note that the word MEMORY must be upper case. Figure 2.2 contains an example of the MEMORY directive, which defines a system that is corresponding to the TMS320C30 microcomputer mode.

The linker has a default memory model that is based on the TMS320C30 architecture. If you do not use the MEMORY directive, the linker uses this default model, which assumes that the full 24-bit address space is present in the system and available for use.

```
/**********************************************************/
/*       Sample command file with MEMORY directive        */
/**********************************************************/

file1.obj
file2.obj                                  /* input files */

-o prog.out                                /* output file */

MEMORY
{
       VECS : origin = 0000000h, length = 00c0h
       ROM  : origin = 00000c0h, length = 1000h
       RAM0 : origin = 0809800h, length = 0400h
       RAM1 : origin = 0809c00h, length = 0400h
}
```

Figure 2.2: An Example of the MEMORY Directive

SECTIONS Directive

The SECTIONS directive tells the linker how to combine sections from the input files into sections in the output module and where to place the output sections in memory.

The SECTIONS directive is specified in a linker command file by the word SECTIONS (upper case), followed by a list of output section specifications enclosed in braces. The general form of the SECTIONS directive is:

```
SECTIONS
{
   section_specification_1
                 .
                 .
                 .
   section_specification_n
}
```

Each section specification defines an output section. The syntax for a section specification is:

```
name [binding or align(n)] :
   {
      input_sections
      assignments
   }  [=fill_value] [ > named memory ]
```

The label `name` names the section in the output module. `Binding` and `align(n)` are optional. The `binding` assigns the section to a specific physical memory address. The `align(n)` specifies that the section should be aligned on an address boundary.

The `input_sections` is a list of input sections that are combined to form the output section. They are enclosed in braces. The `assignments` and `fill_value` are optional and relate to defining the value of symbols or the creation of uninitialized spaces. The `> named memory` tells the linker to allocate the output section into a memory range that was named by the MEMORY directive. Figure 2.3 contains an example of the SECTIONS directive.

This example defines four output sections, `.text`, `.data`, `init`, and `.bss`. The `.text` output section combines the input `.text` sections from `file1.obj` and `file2.obj`. The empty braces, {}, tell the linker to include all input sections with the same name `.text` as the output section. The `> ROM` tells the linker to allocate the output `.text` section into the named

```
/***************************************************/
/*   Sample command file with SECTIONS directive   */
/***************************************************/

file1.obj
file2.obj                            /* input files */

-o prog.out                          /* output file */

SECTIONS
{
   .text  0C0h : { } > ROM

   .data : { file1.obj (.data) }

   init  : { file1.obj (init)
             file2.obj (.data) }

   .bss : { } > RAM0
}
```

Figure 2.3: An Example of the SECTIONS Directive

memory range ROM. An address was specified for the section; this causes the `.text` section to begin at address 0C0h. The `.data` output section contains the `.data` section from `file1.obj`. The `init` section is a named section. It is composed of the input named section `init` in `file1.obj` and the `.data` section from `file2.obj`. The `.bss` output section is composed of the input `.bss` section from `file1.obj` and `file2.obj`. This section will be allocated into the named memory range RAM0.

If you do not supply a SECTIONS directive, the linker uses the following default section specifications:

```
SECTIONS
{
   .text : { }
   .data : { }
   .bss  : { }
}
```

Invoking the Linker

To invoke the linker from the host system, enter the command:

```
LNK30 filename_1, ..., filename_n [-options]
```

The filenames can be object files created by the assembler, linker command files, or achieve libraries. The default extension for all input files is `.obj`; any other extension must be explicitly specified.

The options control linking operations. There are over ten linker options that can be used. For example, you can use the `-e symbol` option to specify an entry executing point for the output module, or use `-o filename` option to name the executable output file.

If you do not use `-o` option to specify an output filename, the linker creates an output file using a default filename `a.out`.

The following example links two files, `file1.obj` and `file2.obj`, and creates an output file named `link.out`.

```
LNK30 file1.obj file2.obj  -o link.out
```

Another way to invoke the linker is to use a command file, which will be discussed in the next section.

```
/**************************************************/
/*         Sample Linker Command File             */
/**************************************************/

file1.obj  file2.obj                  /* input files */
-o prog.out  -m prog.map              /*  operations */

MEMORY
{
     ROM  : origin = 00000c0h, length = 1000h
     RAM0 : origin = 0809800h, length = 0400h
     RAM1 : origin = 0809c00h, length = 0400h
}

SECTIONS
{
    .text : { } ROM
    .data : { } ROM
    .bss  : { } RAM0
    output: { } RAM1
}
```

Figure 2.4: An Example of a Linker Command File

Linker Command File

As described above, you can specify input filenames and options on a command line to invoke the linker. You can also put the filenames and options in a linker command file then invoke the linker from the command line specifying the command file name as an input file.

Linker command files are useful because they simplify the linker command when you invoke the linker often, with the same information. They also allow you to use the MEMORY and SECTIONS directives to customize your program. You can not use these directives from the command line.

Linker command files are ASCII files. They may contain input filenames including other command files, linker options, and linker directives. You can also place comments in a command file by delimiting them with `/*` and `*/`. Figure 2.4 shows a sample linker command file.

To invoke the linker with a command file, enter the LNK30 command and follow it with the name of the command file. Assuming the command file in Fig. 2.4 has the name `linker.cmd`, you invoke the linker by entering:

```
LNK30 linker.cmd
```

2.4 The Evaluation Module and Debugger

The TMS320C30 Evaluation Module (EVM) is a hardware board that lets you execute and debug your applications program using the TMS320C30 C Source Debugger software tool. We will use the EVM and the C Source Debugger to carry out all the experiments in this book. The manufacturer's manuals, TMS320C30 Evaluation Module Technical Reference and TMS320C3x C Source Debugger User's Guide [15, 18], describe the EVM and the debugger in detail. In this section, we give a brief introduction to their features.

2.4.1 The TMS320C30 Evaluation Module

The TMS320C30 EVM is a complete DSP system on a single half-length board that installs in a vacant slot in your PC. When you connect analog input and output (such as a microphone and a speaker) to the system, the EVM becomes a simple real-time signal processing tool. The key features of the EVM board is listed below.

- TMS320C30 33-MFLOP floating-point DSP
- 16k words of zero wait-state SRAM on the primary bus

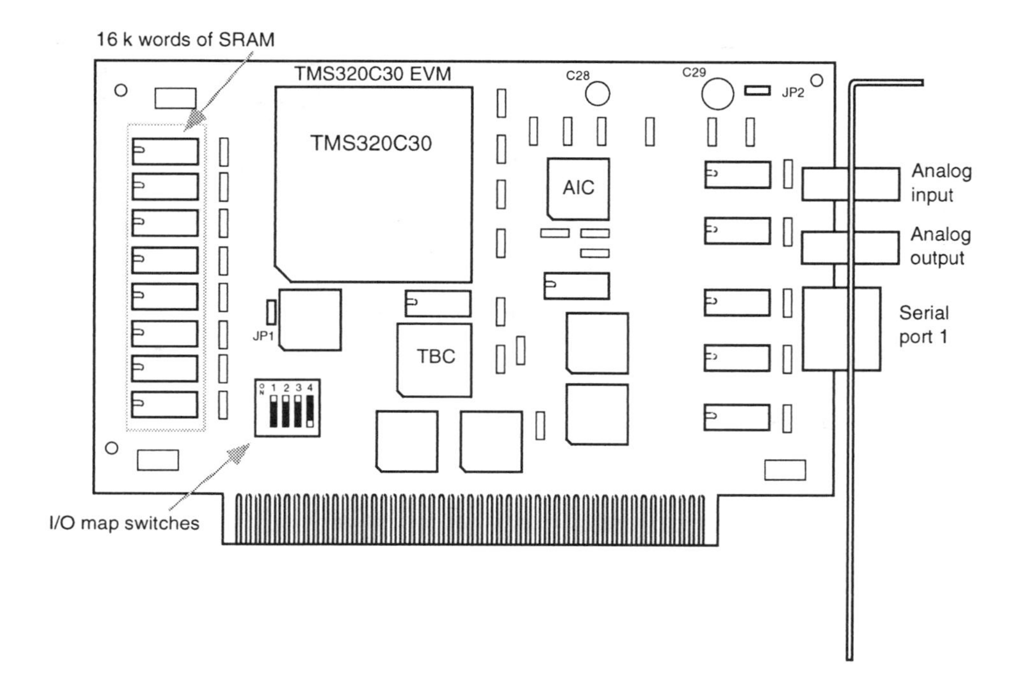

Figure 2.5: TMS320C30 EVM Board with some of its components indicated

- Voice quality analog data acquisition via the Analog Interface Circuit TLC32044
- Embedded emulation support via the 74ACT8990 Test Bus Controller
- 16-bit bi-directional PC host communications port
- External serial port
- Standard RCA jacks for line-level analog input and output
- IBM PC/AT compatible 8-bit half card, mapped to one of four I/O locations

The layout of the EVM board, shown in Fig. 2.5 and Fig. 2.6, shows a basic block diagram of the interconnections of the TMS320C30 EVM.

All program codes for the EVM are loaded through the TMS320C30 emulation port. Once the code has been loaded, the host and the TMS320C30 communicate by means of a shared bi-directional 16-bit register. The register-based interface between the host and TMS320C30 is simple and provides a moderate transfer band-width of approximately 200k bytes per second. This interface has been tailored for use with the TMS320C30's

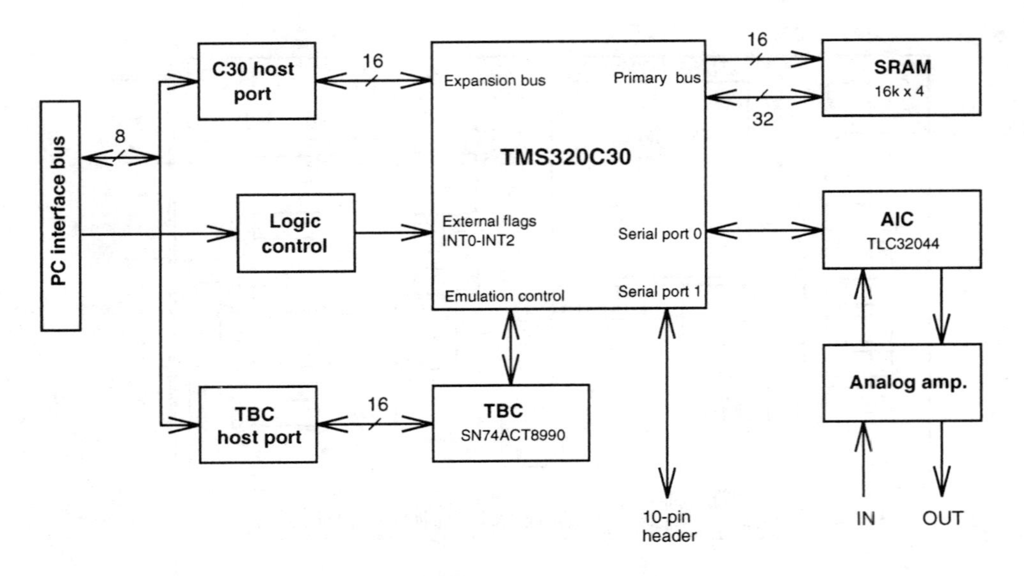

Figure 2.6: TMS320C30 EVM Block Diagram

DMA channel and can be fully interrupt driven.

The EVM's analog section uses a TLC32044 Analog Interface Circuit (AIC) as the analog interface controller. The AIC has an on-chip D/A and A/D conversion with 14 bits of dynamic range, and variable D/A and A/D sampling rates and filtering. It interfaces to the TMS320C30 by means of serial port 0.

The EVM has 16k words of 35-ns SRAM connected directly to the TMS320C30 for storing program or data. This supports zero wait-state memory accesses on the primary bus. The memory is *not* dual-port and can, thus, not be seen directly from the PC side, but data can be transferred to the SRAM memory via the Test Bus Controller (which is what the C Source Debugger, described below, does). Figure 2.7 shows the complete memory map for the TMS320C30 EVM board.

2.4.2 The C Source Debugger

The TMS320C30 C Source Debugger is an interactive window-oriented software interface to the EVM that helps you to develop, test, and refine TMS320C30 C programs and assembly language programs.

The various display windows and the powerful command set of the debugger provides you with the ability to control almost all aspects of the debugging. Through the debugger, you can download your executable object code, display a disassembled version of the code, view the contents of

Address	Contents
0x0	Reset interupt vector
0x1	INT0 interupt vector
0x2	INT1 interupt vector
0x3	INT2 interupt vector
0x4	INT3 interupt vector
0x5	XINT0 interupt vector
0x6	RINT0 interupt vector
0x7	XINT1 interupt vector
0x8	RINT1 interupt vector
0x9	TINT0 interupt vector
0xA	TINT1 interupt vector
0xB	DINT0 interupt vector
0x40 .... 0x7FFFFF	External memory
0x806000 .... 0x807FFF	COM_DATA register (duplicated)
0x808000	DMA global control
0x808004	DMA source address
0x808006	DMA destination address
0x808008	DMA transfer counter
0x808020	Timer 0 global control
0x808024	Timer 0 counter
0x808028	Timer 0 period
0x808030	Timer 1 global control
0x808034	Timer 1 counter
0x808038	Timer 1 period
0x808040	Serial port 0 global control
0x808042	Serial port 0 FSX/DX/CLKX control
0x808043	Serial port 0 FSR/DR/CLKR control
0x808044	Serial port 0 R/X timer control
0x808045	Serial port 0 R/X timer counter
0x808046	Serial port 0 R/X period
0x808048	Serial port 0 data transmit
0x80804C	Serial port 0 data receive
0x808050	Serial port 1 global control
0x808052	Serial port 1 FSX/DX/CLKX control
0x808053	Serial port 1 FSR/DR/CLKR control
0x808054	Serial port 1 R/X timer control
0x808055	Serial port 1 R/X timer counter
0x808056	Serial port 1 R/X period
0x808058	Serial port 1 data transmit
0x80805C	Serial port 1 data receive
0x808060	Expansion bus control
0x808064	Primary bus control
0x809800 .... 0x809BFF	Internal RAM block 0
0x809C00 .... 0x809FFF	Internal RAM block 1

Figure 2.7: Memory map for the TMS320C30 EVM board. All non-specified memory locations should be considered reserved.

the registers and the memory, and run the code. You can also change the data in the registers and the memory locations and display and edit the values of variables, arrays, and structures in hexadecimal, decimal, integer or floating-point formats. You can also execute the code by single-stepping through it or by setting breakpoints and executing between the them.

The C Source Debugger also provides a software-only emulation mode of the TMS320C30. In this mode, all the functionality of the TMS320C30 is performed on the PC, and is hence very slow. For debugging of small code segments on a PC without an EVM board, this can be quite handy. NOTE: The software-only emulation mode does not support interrupt and timer-based functions so some modifications are necessary to debug real-time programs.

In our first experiment, in Chapter 3, we will introduce some basic and often used functions of the debugger. For a full description of the debugger, refer to the corresponding manual [18].

2.4.3 PC Host Communication

The TMS320C30 and the host can communicate via a register-based communications port, which consists of a 16-bit bi-directional register and a Test Bus Controller (TBC) on the EVM board. The 16-bit communication register interfaces the lower eight data bits on the I/O bus of the host PC to the lower 16-bits of the TMS320C30 expansion bus. By reading and writing to this register, data can be transferred from the TMS320C30 to the host PC and vice versa. This allows the PC to interact, on a real-time basis, with the TMS320C30, and to perform non-data intensive tasks such as receiving data from the TMS320C30 for display on the PC screen. Appendix A of this book describes the PC-TMS320C30 data communications in more detail and one of the projects in Chapter 11 shows a complete example of such an interface.

2.5 Experimental Environment

Finally, we collect all the hardware and software required for conducting the experiments in this book and briefly discuss their installation and the DOS environment setup.

2.5.1 The Hardware Setup

The hardware requirements include:

- An IBM PC/AT or 100% compatible PC with a hard-disk and a floppy drive
- TMS320C30 Evaluation Module (EVM board)
- A function generator and an oscilloscope
- A microphone and a speaker (optional)
- Cables with the standard male BNCs and RCA jacks

The "TMS320C30 C Source Debugger User's Guide" [18] contains the step-by-step instructions to install the EVM board. The installation is actually very easy. You just need to remove the cover of your PC and mount the board into one of the 8-bit or 16-bit slots of the PC. The only thing that needs consideration is the setting of the I/O map switches. The default settings that come with the board work in most cases. The analog input and output of the EVM are brought out by the two standard RCA jacks. You can later connect them to the function generator and oscilloscope, or the microphone and speaker. The 10-pin connector on the board connects to the serial port 1 of the TMS320C30.

2.5.2 The Software Setup

The software required includes:

- TMS320C30 Assembly Language Tools Package containing the assembler and the linker (TI release 4.50 or higher)
- TMS320C30 C Source Debugger Software Kit containing the debugger and the associated files (TI release 4.50 or higher)
- TMS320C30 EVM Applications Software Kit containing the EVM communication and some other applications code (TI release 1.10 or higher) (optional)
- A text file editor used to create the source programs

To install the software, copy the packages onto the hard disk of the PC. Usually, we put the different software into different directories in order to maintain them. When doing so, each time you invoke a program, like the assembler, from another directory, you have to specify the path and the name of the directory that contains the assembler. This is very inconvenient and can be avoided by providing the path information in the host system's `autoexec.bat` file. On the other hand, the debugger is designed to find its auxiliary files and other information by using the DOS environment variable D_DIR, D_SRC, and D_OPTIONS. You need to define these variables in the `autoexec.bat` file.

```
PATH=%PATH%;C:\C30TOOLS; C:\EDITOR
SET D_DIR=C:\C30TOOLS
SET D_SRC=C:\C30SRC; C:\EVMASK
SET D_OPTIONS=-p 240
```

Figure 2.8: Sample `autoexec.bat` file to set the DOS environment. The D_OPTIONS setting assumes that the EVM board is installed at I/O address 0x240.

The TMS320C30 C Source Debugger User's Guide contains the instructions and the examples to set up the DOS environment for the debugger. In Fig. 2.8, we give another example of the `autoexec.bat` file that sets up the DOS environment for all the involved software. We assume that the assembly language tools are in the directory C30TOOLS, the C source debugger package is also in C30TOOLS, the EVM application software kit in EVMASK, the text editor in EDITOR, and the directory C30SRC contains the source programs that you create.

2.6 Simulator Environment

Texas Instruments sells a software-only simulator, called `SIM3X`, that emulates the TMS320C30 chip. It allows you to debug your programs without having access to a physical EVM board. The interface to the simulator looks very much like the EVM C source debugger interface and is described in the same manual [18]. Since no hardware exists it is not possible to use interups and, hence, most programs in this manual will not run directly, as they rely on the interupt signal. However, the main portion (for example the FIR filter part) can still be tested through simulation of the I/O by modifying the input register at the beginning of the interupt service routine.

Chapter 3

Sampling

The first step in most digital signal processing systems is the sampling and quantization of an analog input signal (A/D conversion). Similarly, the final step in processing the resulting output is normally the conversion back to an analog signal (D/A conversion). In this chapter, we will consider these operations and their effects on the analog signals.

The purpose of the sampling experiment is three-fold. First, in nearly all applications, we will need code to input and output signal data to and from the processor. Thus, the program code from this section can serve as a template for other experiments, projects, or programs. Second, we will investigate the physical implications of aliasing and A/D amplitude overload (Quantization will be investigated in Chapter 8). Third, a simple sampling program can be used to test our hardware setup. Sometimes, when you are struggling with software and wondering if perhaps your code is correct, but the hardware is faulty, you can run the sampling program to make sure that the hardware does work.

3.1 Sampling Program

In this first experiment, we will begin by giving an example program and explaining its features. Studying example programs is an effective way to learn programming. You should try to understand the program by looking at the instructions and reading the comments. For this sampling program, you will find that most of the code performs a system initialization that is necessary prior to the execution of any DSP algorithm. In this book, we can not describe all the initializations in detail. We will put our emphasis on the setup of the sampling rates because it is what we are most concerned with in this experiment and we will need to change them later. The other initializations will seldom need to be changed. We have tried to give necessary comments in the program to help you understanding the code. For a complete explanation of all initializations, refer to the manuals [14, 15, 19].

3.1.1 System Initialization Routine

For the sake of clarity and the convenience of later use, we make the initialization routine and the main program separate files. Let's first look at the initialization program, which is listed below.

```
*****************************************************************************
*                                                                           *
*                     System Initializations - SYSINIT.ASM                  *
*                     ------------------------------------                  *
*                                                                           *
*   This program initializes the TMS320C30 EVM. It performs the following   *
*   operations:                                                             *
*                                                                           *
*       - Arrange system stack                                              *
*       - Disable/clear all interrupts                                      *
*       - Set the data page pointer                                         *
*       - Enable the cache                                                  *
*       - Initialize the memory ports                                       *
*       - Initialize the AIC                                                *
*       - Enable serial port 0 receive and global interrupts                *
*                                                                           *
*      Jianping Chen,  Henrik Sorensen.                                     *
*      Written Nov. 15, 1991. Modified April 5, 1993. Version 1.0           *
*                                                                           *
*  Do not distribute without permission from authors. Copyright 1993-96     *
*                                                                           *
*****************************************************************************

          .global  start
          .global  evm_init
          .global  p0_addr
          .global  receive0
          .global  stack_addr
          .global  aic_init
          .global  aic_setup
          .global  wt_x_flag

*=================================*
*  Reset and interrupt vectors    *
*=================================*

          .sect    "vectors"
PARMS:
```

```
reset      .word   start              ;reset vector
           .space  5
rint0      .word   receive0           ;serial port 0 recv int. vector
reserved   .space  039h

           .sect   "comdata"

*================================================================*
*  Addresses of various peripherals and memory control registers  *
*================================================================*

mcntlr0    .word   000808064h         ;memory control reg., primary bus
mcntlr1    .word   000808060h         ;memory control reg., expansion bus
t0_ctladdr .word   000808020h         ;Timer 0
p0_addr    .word   000808040h         ;Serial port 0

*=======================================*
*  Control parameters and setup values  *
*=======================================*

t0_ctlinit .word   0C00002C1h         ;set timer 0 as clock out
p0_global  .word   00e970300h         ;serial port 0 global cntl reg value
enbl_sp0_r .word   000000020h         ;serial port 0 receive int enable

WAIT0      .set    0000h              ;memory control reg value, pri. bus
WAIT1      .set    0000h              ;memory control reg value, exp. bus
CACHE      .set    1800h              ;clear and enable cache
ENBL_GIE   .set    2000h              ;global interrupt enable

*=================*
*  System stack  *
*=================*

stack_addr .word   stksec             ;address of stack
stksec     .usect  "stack", stack_size ;reserve space for stack
stack_size .set    040h               ;size of system stack

           .text

* System Initializations

evm_init:  xor     ie,ie              ;disable all interrupts
           xor     if,if              ;clear all interrupt flags
           ldp     PARMS              ;load data page pointer
           ldi     CACHE,st           ;clear and enable cache
           ldi     WAIT0,r0           ;get parallel interface setup value
           ldi     @mcntlr0,ar0       ;get memory control register address
           sti     r0,*ar0            ;set parallel ready
           ldi     WAIT1,r0           ;get i/o interface setup value
           ldi     @mcntlr1,ar0       ;get memory control register address
           sti     r0,*ar0            ;set i/o ready
           call    aic_init           ;routine to initialize AIC
```

```
* Initializations done. Enable receive0 and global interrupts and return to
* main program

          xor     if,if               ;clear out all interrupt flags
          or      @enbl_sp0_r,ie      ;enable receive0 interrupt
          or      ENBL_GIE,st         ;enable global interrupt
          rets                        ;return to main program

************************************************************************
*                                                                      *
*    AIC Initializations                                               *
*                                                                      *
*    Routine to reset and initialize the AIC. It performs the          *
*    following operations:                                             *
*                                                                      *
*       - Set up timer 0 to supply AIC master clock                    *
*       - Reset the AIC                                                *
*       - Initialize the serial ports                                  *
*       - Take AIC out of reset                                        *
*       - Set up the AIC                                               *
*                                                                      *
************************************************************************

* Set up timer 0 to provide AIC master clock

aic_init: ldi     @t0_ctladdr,ar0     ;get address of timer control reg
          ldi     1,r1                ;tclk0 will equal 15MHz/2
          sti     r1,*+ar0(8)         ;set the period register to 1
          ldi     @t0_ctlinit,r1      ;get timer 0 setup value
          sti     r1,*ar0             ;set timer 0 to run in pulse mode

* Reset AIC

          ldi     2,iof               ;set xf0 as output, set xf0 low

* Initialize serial ports

          ldi     @p0_addr,ar0        ;get address of serial port 0
          ldi     111h,r1             ;load port control setup value
          sti     r1,*+ar0(2)         ;initialize transmit port control
          sti     r1,*+ar0(3)         ;initialize receive port control
          ldi     @p0_global,r1       ;get port 0 global cntl setup value
          sti     r1,*ar0             ;initialize port 0 global control

* Clear transmit register

          ldi     0,r1
          sti     r1,*+ar0(8)         ;set transmit data to 0

* Take AIC out of reset
```

```
        rpts    99                 ;wait for 50 timer out clocks to keep
        nop                        ;AIC in reset for some period of time
        ldi     6,iof              ;set xf0 high to take AIC out of reset

* Set up AIC

        call    aic_setup          ;routine to set up the AIC
        rets

***************************************************************************
*                                                                         *
*   AIC Setup                                                             *
*                                                                         *
*   Routine to set up the AIC. It performs the following operations:      *
*                                                                         *
*       - Set sampling rate to 8 KHz                                      *
*       - Set cut-off frequency of antialiasing filter to 3.6 KHz         *
*       - Set AIC control register to enable A/D high pass, disable       *
*         loopback, select primary analog input, synchronous transmit     *
*         and receive, and +/- 1.5-V input, and to insert a (sin(x)/x)    *
*         D/A correction filter.                                          *
*                                                                         *
*   The timing of the writes in these sequences must be synchronized to   *
*   to the AIC data transmission. The subroutine "wt_x_flag" serves this  *
*   purpose.                                                              *
*                                                                         *
***************************************************************************

* Set up lowpass filter cut-off frequency

aic_setup: call wt_x_flag          ;poll for transmit interrupt
        ldi     3,r1               ;secondary transmission word
        sti     r1,*+ar0(8)        ;start a secondary transmission
        call    wt_x_flag          ;poll for transmit interrupt
        ldi     1a34h,r1           ;get counter A setup value
        sti     r1,*+ar0(8)        ;set lowpass filter cut-off freq
        ldi     *+ar0(12),r1       ;void read

* Set up sampling rate

        call    wt_x_flag          ;poll for transmit interrupt
        ldi     3,r1
        sti     r1,*+ar0(8)        ;start a secondary transmission
        call    wt_x_flag
        ldi     4892h,r1           ;get counter B setup value
        sti     r1,*+ar0(8)        ;set sampling rate
        ldi     *+ar0(12),r1       ;void read

* Set up AIC control register

        call    wt_x_flag          ;poll for transmit interrupt
        ldi     3,r1
```

```
           sti     r1,*+ar0(8)          ;start a secondary transmission
           call    wt_x_flag
           ldi     2a7h,r1              ;get control register setup value
           sti     r1,*+ar0(8)          ;setup aic control register
           ldi     *+ar0(12),r1         ;void read
           rets

wt_x_flag: xor     if,if                ;wait for the transmit int. flag
wloop:     tstb    10h,if               ;to be set.
           bz      wloop
           rets

           .end
```

Figure 3.1: Program to Initialize the TMS320C30 EVM

Any line that starts with a "*" or any characters past the ";" are comments. It is generally useful to have the heading contain a brief description of the program.

The next lines, until the `.text` directive, consists of assembler directives that control the assembly process and supply program data. Do not neglect assembler directives. They are as important as instructions. The `.global` directive declares external symbols that are used among modules. The `vectors` section contains reset and interrupt vectors. The reset vector controls the starting point of the program execution. At reset, the program counter is set to `start` (defined in the main sampling program in Fig. 3.3). Likewise, the address loaded at label `rint0` is the serial port 0 receive interrupt vector. Each time an interrupt signal is received from port 0, the program jumps to `receive0` (defined in the main sampling program in Fig. 3.3) to service the interrupt. The next 57 (=039 hex) locations are reserved for other interrupt vectors and traps. The section `comdata` contains the addresses and setup values of the control registers involved in the initialization. The last three lines arrange the system stack. They reserve 64 (=40h) words of space for the stack and associate a symbol with its beginning address.

The actual instructions start after the `.text` directive. The program first enables the cache and sets the memory control registers. Then it calls subroutine `aic_init` to initialize the Analog Interface Controller (AIC). The `aic_init` sets timer 0 to provide a 7.5 MHz AIC master clock and configures serial port 0 as an I/O port to the AIC. It also initializes the AIC by holding it in reset for several cycles to allow it to settle before releasing it. The next step is to set up the sampling rate, the cut-off frequency of the antialiasing lowpass filter, and the AIC control register status. We will

Primary DX Transmission

No following secondary transmission

14-bit twos-complement data	0 0

Secondary transmission to follow

14-bit twos-complement data	1 1

Secondary DX Transmission

15	14	13	12	11	10	9	8	7	6	5	4	3	2	1	0
X X		TA					X X		RA					0 0	
X	TB						X	RB						1 0	
X X X X X X						d9	X	d7	d6	d5	d4	d3	d2	1 1	

Figure 3.2: AIC Transmit Data Word Formats

describe this in detail.

The TMS320C30 chip communicates with the AIC through the memory-mapped register 0x808048h (transmit register) and 0x80804Ch (receive register). In normal or primary operation, input and output samples are transmitted through these registers. In addition, the C30 also downloads configuration commands to set the AIC via the transmit register. This is called a secondary transmission. The AIC data word is 16 bits. The 14 MSBs contain the 14-bit twos-complement data. The two LSBs equal zero for the data received from the AIC. In the transmit section, however, the two lower bits are used as control bits. When these two bits are set to 11 (binary), it signals the AIC that the following transmission will be a secondary transmission. The next data transmitted will contain a configuration word. Figure 3.2 illustrates the AIC transmit data word formats.

The AIC has two registers, A and B, that determine the sampling rate F_s and the cut-off frequency F_{lp} of the antialiasing lowpass filter. The values of A and B are generated according to the following formulas:

$$F_{lp} = F_{clk}/(160A) \tag{3.1}$$

$$F_s = F_{clk}/(2AB) \tag{3.2}$$

The F_{clk} is the master clock frequency derived from the TMS320C30's `timer_0`, which has previously been set to 7.5 MHz. For a sampling rate of 8 kHz, a 3.6 kHz lowpass filter is appropriate. From Eqns. (3.1) and (3.2), we hence see that A and B should be chosen as 13 and 36, respectively. In fact, the AIC has been designed with an inherent antialiasing function by making both F_{lp} and F_s a submultiple of the F_{clk}. With a default value of 36 in register B, we just need to set the A value to achieve the desired sampling rate with a proper lowpass filter. For example, if we want to have a sampling rate around 12 kHz, we need only to set A to 9, since $7500000/(2 * 9 * 36) = 11.6$ kHz. The lowpass filter frequency is now $7500000/(160 * 9) = 5.2$ kHz, which is appropriate for the sampling rate 11.6 kHz. In this experiment, since we want to study aliasing effects, we will set A and B individually.

Sampling frequency	**A** (config word)	**B** (config word)
17.361 kHz (highest)	6 (`0C18h`)	36 (`4892h`)
11.574 kHz	9 (`1224h`)	36 (`4892h`)
8.013 kHz	13 (`1A34h`)	36 (`4892h`)
4.006 kHz	26 (`3468h`)	36 (`4892h`)
3.360 kHz (lowest)	31 (`3E7Ch`)	36 (`4892h`)

Table 3.1: A and B values for typical sampling frequencies

In the configuration word formats shown in Fig. 3.2, TA and RA represents the A register, and TB and RB represents the B register. The T and R fields control the D/A (transmit) and A/D (receive) sections respectively, such that it is possible to run the input and output at different rates. When the A/D and D/A work at the same rate, RA = TA and RB = TB. To use $F_s = 8$ kHz and $F_{lp} = 3.6$ kHz, set RA = TA = 13 and RB = TB = 36. Inserting these A and B values into the appropriate positions in the secondary transmission words in Fig. 3.2 results in the configuration words `1A34h` and `4892h`. Other configuration words are shown in table 3.1.

In addition to the A and B registers, a third set of values is required to set the AIC control register status, which determines the various working modes. Table 3.2 shows the setting when downloading the word `02a7h`.

The AIC also has a highpass filter in the A/D section that cuts off the input frequencies below about 100 Hz. To bypass the highpass filter, turn off the d2 bit. A value of 1 in the d5 bit synchronizes the receive and transmit operations. If you want to set the A/D and D/A to work at different rates, set this bit to zero.

Name	Status	Setting
d2	1	A/D high pass enabled
d3	0	Loopback disable
d4	0	Use primary analog input
d5	1	Synchronous transmit and receive
d6,d7	0,1	+/− 1.5 V input
d9	1	Insert $sin(x)/x$ D/A correction filter

Table 3.2: AIC Control Register Status

After this digression, let us return to the program. We have made the AIC setup a separate subroutine: `aic_setup`. Its first line, `call wt_x_flag`, is a wait for the transmit flag to be set, which means that the AIC is ready to receive data. This instruction will be used before every write to guarantee that the DSP does not write before the AIC is ready, since the DSP is much faster than the AIC (Notice this function is not needed in the sampling program in the next section since synchronization is handled implicitly via the interrupt timing). The next two instructions start a secondary transmission, since writing 3 to the transmit register sets the two LSBs to 11. The following secondary transmission loads the word 1A34h to set the lowpass filter frequency to 3.6 kHz. Likewise, the next two blocks of code transmit the words 4892h and 2a7h to set up the sampling rate (8 kHz) and AIC control register status.

Now, all the necessary initializations are done and the system is ready to run. The program hence enables the serial port 0 receive interrupt and the global interrupt to accept samples from the AIC.

3.1.2 Main Sampling Program

In the previous section, we studied the initialization program. Now let us examine the main sampling program shown in Fig. 3.3.

The first task should always be to initialize the system. However, before calling `evm_init`, the stack pointer must be loaded (i.e., set to the top of the stack) because the `call` instruction will use the stack to store the return address. After the EVM is initialized, the program enters an infinite loop on the `idle` instruction waiting for an interrupt. Each time an interrupt signal is generated, program control is transferred to the `receive0` subroutine to perform the interrupt service. Within the interrupt routine, the program reads a sample from the receive register and then sends it out

```
***************************************************************************
*                                                                         *
*                      Sampling Program - SAMP.ASM                        *
*                      ----------------------------                       *
*                                                                         *
*   This program takes an input sample from the serial port 0 receive     *
*   register and sends it out through the transmit register. It first calls *
*   subroutine "sysinit" to initialize the EVM and sets the system ready to *
*   run. Then it waits in idle loop for AIC receive interrupts. Each time  *
*   an interrupt is generated, it jumps to the interrupt service routine  *
*   to read and write a sample.                                           *
*                                                                         *
*      Jianping Chen,  Henrik Sorensen.                                   *
*      Written in Nov. 15, 1991. Modified in April 9, 1993. Version 1.0   *
*                                                                         *
*  Do not distribute without permission from authors. Copyright 1993-96   *
*                                                                         *
***************************************************************************

* Symbols defined in this program
          .global  start                  ;entry point of the program
          .global  wait_intr              ;wait for interrupt

* Symbols defined in "sysinit"
          .global  evm_init               ;EVM initialization subroutine
          .global  receive0               ;receive0 interrupt routine
          .global  p0_addr                ;serial port 0 address
          .global  stack_addr             ;system stack address

          .text
* Initialize system first
start:    ldi      @stack_addr, sp        ;load stack pointer
          call     evm_init               ;routine to initialize the EVM

* Wait in idle loop for interrupts
wait_intr: idle                           ;wait for interrupts
          br       wait_intr

* Interrupt service routine
receive0:  ldi     @p0_addr,ar0           ;get port address
          ldi      *+ar0(12),r0           ;read a sample
          sti      r0,*+ar0(8)            ;write the sample
          reti                            ;return from interrupt

          .end
```

Figure 3.3: Main Sampling Program

immediately through the transmit register. No processing is done to the data in this program. You can imagine that in the programs for later experiments, we will alter the `receive0` interrupt routine to modify the samples in some appropriate way. After finishing the interrupt service routine, the program control is returned to the main program (which is simply the `idle` instruction in this case) via the `reti` instruction.

3.1.3 Linker Command File

The linker combines separate object files to produce the final executable program. In most cases, a linker command file must be used. Fig. 3.4 shows an example of the linker command file for the sampling program. In the command file, you specify the input object files and output file name, and define and allocate the memory. It is useful to compare the memory layout of `samp.cmd` below with the memory map in Fig. 2.7. This step is important not just at the compilation stage, but much earlier in the programming task. Often, it is necessary to consider how the memory will be used, even before the programming begins.

```
/**************************************************************************/
/*                                                                        */
/*         Linker Command File for Sampling Program - SAMP.CMD           */
/*        ----------------------------------------------------            */
/*                                                                        */
/*   This linker command file links the assembler object file samp.obj  */
/*   and sysinit.obj to produce the output file samp.out that can be     */
/*   loaded into the EVM to execute.                                      */
/*                                                                        */
/*     Jianping Chen,  Henrik Sorensen.                                   */
/*   Written in Nov. 15, 1991. Modified in April 9, 1993. Version 1.0    */
/*                                                                        */
/*   Do not distribute without permission from authors.                  */
/*   Copyright 1993-96                                                    */
/*                                                                        */
/**************************************************************************/

samp.obj                        /* input files */
sysinit.obj

-e start                        /* label  for entry point of code */
-o samp.out                     /* output file name */

MEMORY                          /* define memory ranges */
{
    INT_V : origin = 0x000000, length = 0x40
    SRAM  : origin = 0x000040, length = 0x3FC0
    RAM0  : origin = 0x809800, length = 0x400
```

```
    RAM1   : origin = 0x809C00, length = 0x400
}

SECTIONS                             /* allocate program sections to memory */
{
         vectors: {} > INT_V       /* interrupt vectors in INT_V */
         comdata: {} > SRAM        /* section comdata in SRAM */
        .text    : {} > SRAM        /* section .text in SRAM */
        .bss     : {} > RAM0        /* section .bss in RAM0 */
         stack   : {} > RAM1        /* section stack in RAM1 */
}
```

Figure 3.4: Linker Command File

The `samp.obj` and `sysinit.obj` are the input object files created by the assembler. The linker option `-e` defines the symbol `start` as the entry point from which the program starts executing. The MEMORY directive specifies memory ranges. Each memory range has a name, a starting address (`origin`), and a length. The SECTIONS directive combines input sections to form output sections and allocates them into specific memory areas.

Now that we have gone through all the code required for the sampling program, it is time to assemble, link, and execute the program. Through the following experiments, you will become familiar with the procedure of program assembling and linking, and with the use of the EVM debugger. At the same time, the experiment will check your hardware setup.

3.1.4 Experiment 3A: Executing Sampling Program

Before beginning the experiment, we assume that you have properly set up the hardware system and the software environment according to the previous chapter and the manufacturer's manual [17, 15]. The assembler, linker, debugger, and other required files should all be in place.

Follow the instructions on the included floppy and install the files onto your harddisk. Locate the sampling programs `samp.asm`, `sysinit.asm`, and `samp.cmd` in the newly created directory.

The first step is to assemble and link the program. At the DOS prompt, enter the following command:

```
ASM30 samp.asm
```

An object file samp.obj should be created with no error message. Repeat the command to create sysinit.obj. Then, link the object files by typing the command:

```
LNK30 samp.cmd
```

The linker should produce the output file `samp.out` without any error. This `samp.out` now can be loaded into the EVM to execute.

Invoke the debugger and load the `samp.out` by entering the command:

```
EVM30 samp.out
```

Alternatively, you could first open the debugger by typing EVM30 and then load the program. The loading can be done by either entering `LOAD samp.out` in the COMMAND window or by pressing the `Alt-L` keys as prompted at the top line on the screen. These commands are useful when you are already in the debugger and want to reload your program or switch to another program.

Now the debugger windows are displayed on the screen. You will first see a message in the COMMAND window stating that the program `samp.out` is loaded. Shown in the DISASSEMBLY window is the assembly language version of `samp.out`. The entry point line is highlighted (at address `start`), indicating that the PC is currently pointing to this instruction. You can confirm this by looking at the value of the PC counter in the CPU window. The MEMORY window initially displays memory contents starting at address 0.

Before running the program, check again to make sure that the function generator and the oscilloscope are properly connected and set. Reset the processor by entering the command:

```
RESET
```

Run the program by entering the command:

```
RUN
```

If the entire system is functioning properly, you should see that the input signal passes through the EVM and is displayed on the oscilloscope. If you have connected a microphone and a speaker, any sound generated at the microphone will be reproduced by the speaker. Apply a sinusoidal input signal and adjust the frequency from 50 Hz up to 4 kHz. Observe the output via an oscilloscope and notice that for most frequencies it is a perfect replica of the input signal except for some gain and a short delay (Why is there a delay and why is it frequency dependent?). Determine the gain of the system (i.e., ratio of output amplitude to input amplitude) as a function of frequency.

Pressing the ESC key, will halt the processor and disrupt the signal path. Entering the RUN command resumes the processor. If it does not work, apply the RESET command before the RUN. Try halting and resuming the program a few times.

Now use the program to gain some practice with the debugger.

- Change the active window by pressing the F6 key, by using the command "WIN [window name]", or by clicking with the left mouse button in the desired window.
- Use the command "MEM [address]" to modify the memory window to display the same object code that is shown in the disassembly window.
- Use the command DASM or "ADDR [address]" to change the starting line of the code displayed in the disassembly window.
- Try the arrow keys, the Pg-Up and Pg-Dn keys, and the left mouse button in the different windows to see how they function.
- Single-step through the program by pressing the F8 key or by typing the command STEP. Observe the flow of the program and the change of the contents of the registers in the CPU window.
- Use the command "GO [expression]" to run the program up to a specific point. The "expression" may be an address or a label.
- Set a breakpoint at the `sti` instruction in the interupt service routine using the F9 key, by left-clicking on the instruction, or by using the BREAK menu. Type RESET and RUN to execute until the breakpoint. Notice that, although it is possible to continue executing the program using F5 or F8, no more interupts will occur, which means the program will become indefinitely stuck in the `IDLE` instruction.

Practice the steps above until you are comfortable making the debugger run or step through a program, and display the content of any register or memory location. These are the most often used operations when you are debugging programs.

Slowly increase the amplitude of a 1 kHz sinusoidal input voltage until the output of the EVM board start to clip (i.e., start to turn into a square wave). What is the EVM input voltage when this starts to occur? What is the EVM output voltage when this occurs? Determine whether it is the A/D or the D/A converter that is clipping at this input level.

Lower the voltage again and change your input to a squarewave and sweep its frequency from 50 Hz to 4 kHz. Explain what you see on the oscilloscope.

3.2 Aliasing Effects

From the theory of sampling, we know that if an analog signal is sampled at F_s samples per second then the resulting discrete signal has frequencies up to $F_s/2$ Hz. Any frequency in the analog input above $F_s/2$ is aliased into these lower frequencies. We will verify this fact in the following experiment. As we have discussed in Section 3.1.1, the sampling rate F_s and the lowpass filter frequency F_{lp} are determined by the values in the A and B registers of the AIC. By changing the value B, we can change the F_s to cause aliasing.

3.2.1 Experiment 3B: Study of Aliasing Effects

This experiment is a continuation of Experiment 3A. We assume that you have just completed Experiment 3A and that the program `samp.out` is running on the debugger. At the moment, the system is working at an 8 kHz sampling rate with a 3.6 kHz lowpass filter.

Aliasing from Insufficient Sample Rate

Aliasing will occur if we lower the sampling rate to 4.6 kHz with F_{lp} at 3.6 kHz unchanged. From Eqn. (3.2), B should be 63 for $F_s = 4.6$ kHz. The corresponding configuration word is `07EFE`h. Now we need to change the value `04892`h to `07EFE`h in the program `samp.out`. One way to do this is to go back and edit the source file, `sysinit.asm`, and then reassemble and link, etc. This is not the fastest way. We can actually make the change directly in the debugger!

Halt the processor. From the DISASSEMBLY window, locate the instruction `LDI 18578, R1` (note that `18578` is the decimal version of the hexadecimal number `04892`h) and observe that the corresponding object code is `08614892`. Now note the address of this instruction and locate the same address in the MEMORY window (You can also look for the same object code, obviously). With the practice from the previous experiment, you should be able to carry out these operations. Now we are going to change the object code `08614892`h to `08617EFE`h directly on the debugger. Make the MEMORY window the active window. Move the cursor to the field `08614892`. Press the F9 key to highlight the field. Then, type `08617EFE`. You will see it overwrite the original value. After you finish the typing, press the ENTER key. The content of this location is now replaced with the new value `08617EFE`h. Notice that the instruction in the DISASSEMBLY window has changed accordingly. The next memory location is now highlighted. Press the ESC or the F9 key to turn off the highlight. If you make a mistake or change your mind during the editing, press the ESC or the F9 key to reset the field to its original value.

You have just succesfully edited data in a memory location (same way for a register). This is a very useful skill for debugging. After making the change, we now can execute the program and observe aliasing. Reset the processor and run.

Adjust the input sinusoidal signal frequency from 100 Hz up to 4 kHz and watch the output waveform on the oscilloscope. Find out the starting frequency at which aliasing occurs. Verify it theoretically. By carefully adjusting the input frequency around 2.3 kHz, you will observe a frequency at which the aliased output is again a single sinewave with the same frequency as the input. Why? Change the input frequency below and above 2.3 kHz and observe the output.

At some input frequencies, the output resembles an amplitude modulated (AM) signal. Adjust the input frequency until this happens and explain why it occurs.

Removing Aliasing by Bandwidth Limitation

To avoid the aliasing at $F_s = 4.6$ kHz, the lowpass filter should have a cutoff frequency of about 2 kHz. This requires a value of 23 for A in Eqn. (3.1). The corresponding transmit word is `02E5Ch`. Find out the instruction that loads the word A and change it to the new value on the debugger. Can we now run the program? No! The value of B was changed in the previous section to cause aliasing; and we must recover it. As you can see from Eqn. (3.2), the sampling rate is affected by both A and B. A=23 and B=63 will make the sampling rate 2.6 kHz. Change B back to 36. Then run the program. Now the system is working at the sampling rate of 4.6 kHz with a 2 kHz lowpass filter. No more aliasing should be seen.

We have mentioned in Section 3.1.1 that the AIC has an inherent antialiasing function, and by changing only A we can get a desired sampling rate with an appropriate lowpass filter. We have actually just confirmed this by changing B back to 36. Verify this again by replacing the word `02E5Ch` with `01224h` (A=9). The system will be working at $F_s = 11.6$ kHz with $F_{lp} = 5.2$ kHz.

Finally, increase the input amplitude until the output waveform starts to degenerate. What are the input and output amplitudes when this occurs and what can you deduce from that?

3.2.2 Experiment 3C: Further Modification of Sampling Program

In this experiment we will initially use a 13 kHz sampling rate with no aliasing corresponding to using $A = 8$ and $B = 36$ in Experiment 3A.

3C.1 Determine AIC set-up words and run the sampling program with these new values. Use the function generator and oscilloscope to obtain the frequency response of the system.

3C.2 Rather than writing the input data directly to the D/A converter, set 2 out of every 3 samples to zero before writing them out to the D/A. Measure the new frequency response and explain what happens.

3C.3 Continue to "zero out" 2 out of every three samples as in 3C.1, but now modify the output "sampling" frequency to $\frac{13}{3}$ kHz, while leaving the input sampling frequency at 13 kHz (to do this you must clear the d5 bit in the AIC control word in Table 3.2). Find frequency response of system and compare to frequency response obtained in 3C.1. Explain what happens.

In these two experiments, we are modifying the original sampling program (i.e., 8 kHz sampling rate with $A = 13, B = 36$).

3C.4 Rather than copying the input data directly to the D/A converter, multiply *every other* sample by (-1). Use a slow sweeping sinusoidal input and observe the output. Explain your observations from a theoretical viewpoint.

3C.5 This idea was used in early speech encryption devices. Listen to your output by attaching a microphone to the A/D input and a speaker to the D/A output. Is the speech intelligible? How do you "decrypt" the speech again?

Chapter 4

Waveform Generation

A waveform (or function) generator is an integrated part of many applications implemented on digital signal processors. In this chapter, we discuss the implementation of a squarewave, sinewave, and random number generator using the TMS320C30.

4.1 Squarewave Generator

4.1.1 Squarewave Generator

A squarewave assumes a constant amplitude, $+A$ or $-A$, alternating with some fixed rate. Because of the sampled nature of the signal, each of these constant levels actually consists of several samples. A typical sampled squarewave might look like:

$$x[n] = \{\ldots, A, A, A, -A, -A, -A, A, A, A, \ldots\} \tag{4.1}$$

The number of samples in a row with constant amplitude (i.e. the number of samples with amplitude A) is determined by the D/A conversion frequency F_s and by the desired squarewave frequency.

4.1.2 Implementation of Squarewave Generator

The implementation of a squarewave generator can be done very similar to the play-back part of the program from experiment 3A. The program should be modified to output either A or $-A$, as appropriate, instead of the input samples to the D/A converter. Hence, the interrupt service routine from the sampling program in Fig. 3.3 should now look something like Fig. 4.1

NOTICE: Do not remove the second LDI statement after receive0, which reads from the A/D. Unless we want to change the AIC setup, this dummy read is needed to clear the interrupt flags. We shall see later how to get rid of that. Also observe the statement lsh 2,r0 towards the end of the program. This instruction is needed to convert the output into the format specified in Fig. 3.2,(i.e., 14 data bits followed by two zeros).

```
******************************************************************************
*                                                                            *
*                    Squarewave Program - SQUARE.ASM                         *
*                    -------------------------------                         *
*                                                                            *
*   This program generates a squarewave on D/A port. The program structure  *
*   is very similar to the sampling program, discussed earlier               *
*                                                                            *
*      Henrik Sorensen.                                                      *
*      Written Nov. 17, 1993. Modified in April 9, 1994. Version 1.0         *
*                                                                            *
*  Do not distribute without permission from authors. Copyright 1993-96      *
*                                                                            *
******************************************************************************

          .global  start                  ;entry point of the program
          .global  wait_intr              ;wait for interrupt

* Symbols defined in "sysinit"
          .global  evm_init               ;EVM initialization subroutine
          .global  receive0               ;receive0 interrupt routine
          .global  p0_addr                ;serial port 0 address
          .global  stack_addr             ;system stack address

          .text
* Initialize system first
start:    ldi     @stack_addr, sp         ;load stack pointer
          call    evm_init                ;routine to initialize the EVM

* Initialize  global variables
          ldi     100,ar6                 ;set MAX=100
          ldi     0,ar7                   ;set CNT=0
          ldf     50.0,r7                 ;set AMP=50.0
          mpyf    r7,r7                   ;set AMP=2500.0

* Wait in idle loop for interrupts
wait_intr: idle                           ;wait for interrupts
          br      wait_intr

* Interrupt service routine
receive0: ldi    @p0_addr,ar0             ;get port address
          ldi    *+ar0(12),r0             ;read input (dummy operation
                                          ;needed to clear interrupt)
```

```
* Update output counter
          addi  1,ar7                   ;CNT=CNT+1

* If max reached,
* reset counter and flip amplitude
          cmpi  ar6,ar7                 ;if CNT==MAX {
          bnz   out                     ;
          ldi   0,ar7                   ;   CNT = 0
          negf  r7,r7                   ;   AMP = AMP * (-1) }
                                        ;endif
* Output amplitude
out:      ldf   r7,r0                   ;R0 = AMP

* Convert output to 14 bit integer with two trailing zeros
          fix   r0,r0                   ;convert from float to integer
          lsh   2,r0                    ;change to xxxxxxxxxxxxxx00 format
          sti   r0,*+ar0(8)             ;send output to D/A
          reti

          .end
```

Figure 4.1: Interrupt service routine for squarewave

4.1.3 Experiment 4A: Squarewave Generator

4A.1 Run the program in Fig. 4.1 and measure the frequency on the output (It uses `MAX=100` and $F_s = 8$ kHz). Explain your result.(NOTE: To create `square.out`, you need to create `square.cmd`, which can be done trivially by using `samp.cmd` as a template and simply changing the *samp* name to *square* in two locations.)

4A.2 Change A to 4 and see what happens. Use other values of A but note that A should always be a multiple of 4. Why?

4A.3 Assuming a 100 Hz frequency is desired. Modify the program in Fig. 4.1 to achieve this.

4A.4 Gradually increase your squarewave frequency to 4kHz. Comment on the quality of the waveform as the frequency increases. Explain the effects you see.

4A.5 Modify your program to produce a triangular or sawtooth waveform instead of the squarewave.

4.2 Sinewave Generator

4.2.1 Sinewave Generator

A second order difference equation

$$y[n] = B_1 * x[n-1] + A_1 * y[n-1] + A_0 * y[n-2] \tag{4.2}$$

where $B_1 = 1, A_0 = -1$, $A_1 = 2\cos(\theta)$, and $x[n]$ is the delta function:

$$x[n] = \begin{cases} 1 & n = 0 \\ 0 & \text{otherwise} \end{cases} \tag{4.3}$$

will have a sinusoidal solution. If we, hence, calculate $y[n]$ from this equation recursively, we will get an oscillatory output.

The oscillation results from a pair of poles on the unit circle when the coefficient A_0 is set to -1. The angle θ is determined by the middle coefficient A_1 and the poles are at an angle θ if $A_1 = 2\cos(\theta)$. The oscillation frequency is related to θ as $F = (\theta/2\pi)F_s$, where F_s is the sampling frequency. Thus, the frequency is given by the formula:

$$F = \frac{F_s}{2 * \pi} \arccos(A_1/2) \tag{4.4}$$

From this it is clear that for any sampling frequency, A_1 must be taken on a range between -2 and $+2$. The amplitude of the digital oscillation is $1/\sin(\theta)$ times the starting impulse, $x[n]$.

A sinewave output can, hence, be obtained by computing a special case of the second order IIR digital filter, which will be discussed in Chapter 6. Figure 4.2 shows the structure of the digital sinewave oscillator, and Fig. 4.3 shows a simple Matlab program that implements this algorithm.

4.2.2 Implementation of a Sinewave Generator

The TMS320C30 has many features that facilitate the implementation of Eqn. (4.4), since IIR filtering is an important component of many signal processing tasks. First, it has floating-point arithmetic capabilities. When using the floating-point operations, we do not need to worry about the overflow (normally) and do, hence, not need to scale the algorithm, as we would have using fixed-point operations. Because of the recursive nature of the solution method, accuracy is crucial to prevent error accumulation, so one should use the floating-point instructions to implement the sinewave generator.

The crucial operations for implementing the difference equation in Eqn. (4.2) are the coefficient multiplication and the product accumulation. The

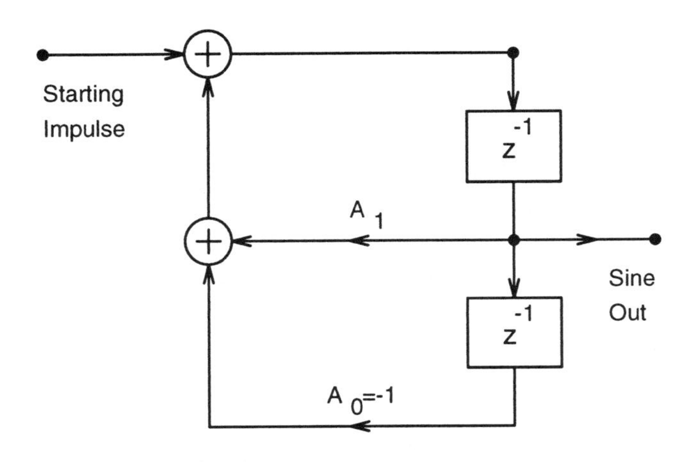

Figure 4.2: Digital sinewave oscillator

```
function [Y] = singen(N,F,Fs)
%SINGEN generates N samples of sinewave.
% [Y]=SINGEN(N,F,FS) generates N samples of sinewave with
%   frequency F at sampling frequency Fs.
%   Example:
%     [Y]=singen(100,1000,8000)
%

%NOTICE: Not written for optimal Matlab performance,
%but to illustrate C30 implementation.

% Henrik Sorensen, Mar. 6, 1992

A1=2*cos(2*pi*F/Fs);   % Constant which controls frequency of sinewave
y1=1;                  % y[1]=1 (from delta function)
y2=0;                  % y[0]=0
for i=1:N
   yn=A1*y1-y2;        % y[n]= delta[n]+A1*y[n-1]+A0*y[n-2]
   y2=y1;              % y[n-2]=y[n-1]
   y1=yn;              % y[n-1]=y[n]
   Y(i)=yn;            % Put y[n] to output
end
```

Figure 4.3: Matlab program C.15 for computing digital sinewave oscillator

hardware multiplier of the TMS320C30 makes it possible to perform the multiplication in only one instruction (cycle). By using a parallel instruction, the coefficient multiplication and the product addition can even be performed simultaneously.

The key point of the implementation of the difference equation is the shift operation of $y[n]$. The latest two output samples need to be stored and kept updated. The TMS320C30 has a circular addressing mode that can efficiently implement this shifting buffer. We will discuss the circular addressing in detail in the filter section later. In this experiment, you may use any method to implement the shift operation.

While we need to use floating-point arithmetic to implement the difference equation, Eqn. (4.2), the output data must be converted to an integer before being transmitted to the Analog-Interface-Chip (AIC). (Conversions between floating-point and integer representations can be done using the `FLOAT` and `FIX` instructions of the TMS320C30 [19].) After being converted to an integer, the output data must be left-shifted two bits to set the normal AIC transmission control.

Some care is necessary in choosing the gain, B_1, of the initial impulse $x[n]$. Since the output data are finally converted to integers to be transmitted to the AIC, you need to choose a gain small enough so that the output data do not exceed the 14-bit two's-complement range of the AIC, and large enough so that the sample values do not suffer excessive quantization effects.

4.2.3 Experiment 4B: Digital Sinewave Generator

4B.1 Verify the solution to the difference equation, Eqn. (4.2)

4B.2 Run the Matlab program to verify the sine-wave generator. Plot the output for 100 output values for different values of F and F_s.

4B.3 Write a program to implement the sinewave generator on the EVM board and output the sinewave to an oscilloscope. Measure the output frequency and compare with your results from 1.

4B.4 Change the coefficient A_1 to produce different frequencies of sinewaves. Compare with theoretical solutions.

4B.5 Change gain B_1 of the input impulse and the sampling rate to see their effects. Explain what happens when B_1 gets "large."

4.3 Random Number Generator

4.3.1 Random Number Generator

The method often used to generate a random number on a digital computer is the computation of a recursive function that approximates a random number. Many such functions have been proposed. The method used in this experiment to generate a pseudo uniform random number is based on the linear congruence method, which uses the following recursion:

$$x[n+1] = 65539 * x[n] \bmod 2^{31} \tag{4.5}$$

If $x[0]$, the initial value of $x[n]$, is chosen as any odd integer with nine or less digits, all 2^{31} possible integers are generated before the output sequence repeats. The sequence, hence, approximates an ensemble of a random variable uniformly distributed between 0 and $2^{31}-1$. A uniformly distributed, floating-point random number, X, in the range of 0 to 1.0 can then be obtained by scaling the integer random number with the following equation:

$$X = x[n+1] * 4.656612875245797 \times 10^{-10} \tag{4.6}$$

An approximation to a gaussian distributed random number Y can be found by computing a sequence of uniform random numbers using the formula:

$$Y = \frac{X_1 + X_2 + \ldots + X_K - \frac{K}{2}}{\sqrt{\frac{K}{12}}} \tag{4.7}$$

where X_i are uniformly distributed random numbers, such that $0 \leq X_i \leq 1$, as in Eqn. (4.6). K is the number of values of X_i used in the summation. Y approaches a true gaussian distribution asymptotically as K approaches infinity.

To acquire the desired mean and standard deviation, scale the random variable Y:

$$Y' = Y * \sigma + \mu \tag{4.8}$$

where Y' is the desired gaussian distributed random variable with standard deviation σ and mean μ.

4.3.2 Implementation of a Random Number Generator

As can be seen from the above discussion, the implementation of the random noise generator can be divided into three steps: First, implementation of Eqn. (4.5) to generate the uniform random integer numbers; Second, computation of Eqn. (4.6) to get the uniform random floating-point number; Third, the generation of the normally distributed random number by implementing Eqns. (4.7) and (4.8).

The implementation of Eqn. (4.5) requires fixed-point arithmetic. Since the TMS320C30 has a hardware multiplier, we might first consider using `MPYI` instruction to perform the multiplication $x[n]*65539$. However, notice that the integer multiplication of the TMS320C30 assumes a 24-bit input and a 32-bit output, while Eqn. (4.5) requires both 32-bit input and output data. Hence, to prevent unwanted periodicities in the random sequences, one should not use the hardware multiplier directly. Instead, the multiplication can be implemented with shift and add operations or by breaking the multiplication up into smaller parts that can be performed without loss on the hardware multiplier. The Arithmetic Logic Unit (ALU) that carries out the shift and add instructions ensures 32-bit integer operations on both input and output data.

To implement the mod 2^{31} operation, the following programming scheme is usually used on a 32-bit general-purpose computers:

$$\begin{array}{ll} \text{Let} & x'[n+1] = x[n] * 65539 \\ \text{If} & x'[n+1] \geq 0 \\ & \text{then } x[n+1] = x'[n+1] \\ \text{Otherwise} & \\ & x[n+1] = x'[n+1] - 2^{31} \end{array} \quad (4.9)$$

We may still use this method on the TMS320C30, but it is not well suited and takes quite a few instructions. A better way is to use the `AND` instruction, which can implement the modulo operation in a single instruction. What should you `AND` with to achieve this?

To implement the accumulation of the X_i in Eqn. (4.7), a loop operation is required. Beside the branch instruction that can be used to implement loops, the TMS320C30 has a block-repeat construct that efficiently implements a loop. With a one-time setup of the `RC` register, the `RPTB` instruction can repeat a block of instructions a desired number (content of `RC` register) of times with no overhead.

The larger the K in Eqn. (4.7), the better an approximation to a gaussian random variable, but the more execution time is required. Therefore, if you choose a big K, you should be careful that your code runs fast enough to calculate a new $y'[n]$ before the arrival of the next interrupt. Generally, there should not be any problem with the speed if you select K small. For example, you can set K to 12 or 48 for convenience.

4.3.3 Experiment 4C: Random Noise Generator

4C.1 Two Matlab programs that computes the random numbers and gaussian variables, as described above, are given in Appendix C. Run the programs and verify the mean and variance of the gaussian random noise for different parameter choices.

4C.2 According to the Eqn. (4.5), (4.6), (4.7), and (4.8), write a program to generate gaussian distributed numbers on the TMS320C30 EVM.

4C.3 Connect the EVM to an oscilloscope and a speaker, respectively, to observe and listen to the output. Comment on the results.

4C.4 Examine the frequency spectrum of the gaussian number generator using a spectrum analyzer. How "white" is your noise? What are some of the limiting factors?

Chapter 5

FIR Filter Implementations

5.1 FIR Filters

Digital filters are a common ingredient of many digital signal processing systems. There are two types of digital filters namely Finite Impulse Response (FIR) and Infinite Impulse Response (IIR). In this chapter, we describe the implementation of FIR filters using the TMS320C30.

An FIR filter has several design advantages over an IIR filter. It is always stable and realizable, and can always be designed to have exact linear phase. The design process, [11], is also less complicated, but the FIR filter does require many more taps than the IIR filter to meet a given set of filter specifications.

An FIR filter is described by the difference equation (or convolution formula)

$$y[n] = h[0]x[n] + h[1]x[n-1] + \ldots + h[N-1]x[n-(N-1)] \qquad (5.1)$$

where $h[n], n = 0, \ldots, N-1$ is the impulse response of the FIR filter. The structure of the FIR filter in its direct form (or tapped delay line) realization is shown in Fig. 5.1.

At any given time, n, the output $y[n]$ is a convolution sum of the impulse response, $h[n]$, and the filter input, $x[n]$. The calculation involves multiplying each of the delayed input samples $x[n-k]$ by a "tap weight" $h[k]$ and adding all of the products. After one sum is obtained, the time index is incremented, and the delay line must be updated by shifting all samples in time, and a new input is made available as $x[n]$.

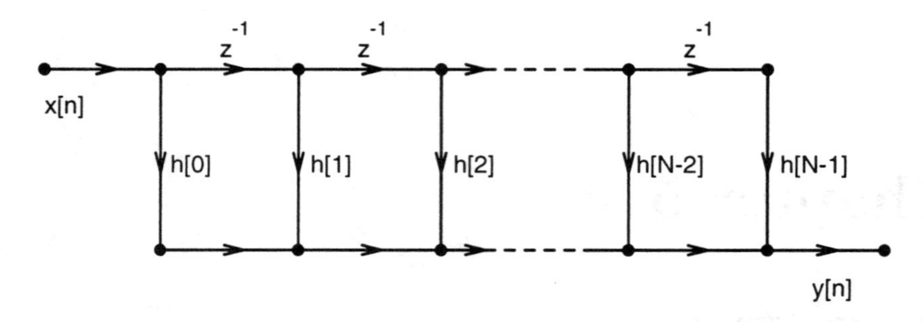

Figure 5.1: Direct form implementation of an FIR Filter

FIR filter design techniques are used to determine the actual tap weights or impulse response $h[k]$ that will adequately approximate a desired filter frequency response or other desired specifications. There are many methods for FIR filter design. Three well-known approaches are the window design method, the discrete-least-square design method, and the Parks-McClellan algorithm. Since the design of FIR filters is described in most texts on digital signal processing including [8, 9, 11], and useful design programs are available for Matlab [7], we do not discuss them here. We will concentrate on the issue of FIR filter implementation on the TMS320C30. From an implementation point of view, the difference between different design methods is just a different set of the filter coefficients.

5.2 Implementation of FIR Filters

Three features of the TMS320C30 that facilitate the implementation of FIR filters are its floating-point capability, its parallel multiply/add instructions, and its circular addressing mode. Using floating-point arithmetic, we can directly utilize the filter coefficients obtained from the filter design programs, whereas a fixed-point implementation would involve scaling the coefficients. The multiply/addition instructions permit the execution of a multiplication and an addition in a single machine cycle; thus permitting the calculation of one filter tap per machine cycle. The circular addressing mode is an efficient way of implementing the data shifts via a circular buffer.

As can be seen from Eqn. (5.1), the arithmetic of an FIR filter involves filter coefficient multiplications and product accumulations (additions). The parallel multiply/add instruction is optimized for these operations. Each coefficient multiplication can be performed in parallel with the running accumulation of the product produced in the previous step. Thus, only N instructions (cycles) are needed to implement the N multiplications

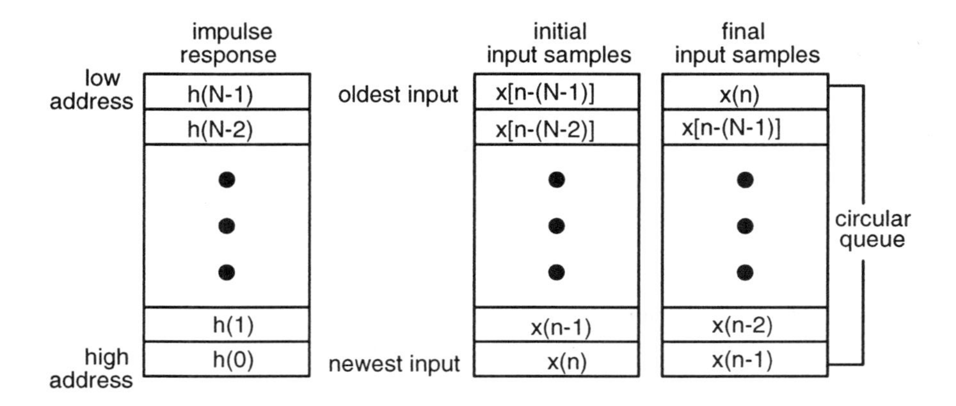

Figure 5.2: FIR filter implementation with cyclic buffer

and N additions needed to compute a length N FIR filter.

Another important part of an FIR filter is the time shift of the input samples $x[n]$. At any given time index, n, the current input $x[n]$, and the previous $N-1$ input samples $x[n-1], \ldots, x[n-(N-1)]$ are needed to compute $y[n]$ and must be stored. Thus, a memory buffer with at least N elements is required to contain the input samples, $x[n], \ldots, x[n-(N-1)]$. When a new sample arrives, it becomes the current $x[n]$. Before it can be placed in the first physical location of the buffer, the other N old samples must all be shifted downward one location. The previous $x[n]$ now becomes $x[n-1]$, the previous $x[n-1]$ becomes $x[n-2]$, etc. and the oldest sample $x[n-(N-1)]$ is pushed out of the buffer since it is no longer needed. Implementing the shift operations in this way (let's call this method "shifting in memory"), we have to move N samples from one memory location to the next. This requires at least N extra instructions.

The circular addressing mode of the TMS320C30 provides a very efficient way to implement the shift operation. Instead of actually moving the data around in the memory, an address pointer is used that circularly moves within the $x[n]$ buffer. When a new sample $x[n]$ is received, it is not always put in the first location of the buffer. It is stored in the same location as the previous oldest sample $x[n-(N-1)]$, which is no longer needed, so it can be overwritten. From this location, which now contains $x[n]$, the pointer successively points to the next location where the current $x[n-1]$ is, and next to $x[n-2]$, etc. When it reaches the physical end of the buffer, it is automatically returned to the beginning to point on the first sample (which is not $x[n]$ in general). Figure 5.2 shows the circular buffer.

The circular addressing is a special form of the indirect addressing mode.

Since the TMS320C30 has two ARAUs (Auxiliary Register Arithmetic Units) that can calculate data addresses independently of the multiplier and ALU, the shift operations can be performed at the same time the coefficients are being multiplied and the products are being added and, hence, no extra cycles are required.

While the circular addressing mode can use variable step sizes; in this application, the step is fixed at 1. The block-size register `BK` controls the length of the circular buffer and should be initialized to N. For the circular addressing to work properly, the starting address of the $x[n]$ buffer must be a multiple of the smallest power of 2 that is greater than or equal to N. For example, $N = 14$ requires the first address for $x[n]$ to be a multiple of 2^4 (the lower 4 bits of the beginning address should be zero). This is needed by the circular addressing logic that is based on the value in the `BK` register and the content of the auxiliary register `ARn` that is used as the pointer.

Another instruction that is useful for the implementation of FIR filters is the single-repeat, `RPTS`, instruction. If the `RC` register is set to $N - 1$, the `RPTS` instruction will repeat one single instruction N times without any overhead. Thus, by using the parallel multiply/add instruction (in combination with the circular addressing mode), the N coefficient multiplications and product additions can be executed in N machine cycles.

Remember, as was done in the experiments on waveform generation, the output data must be properly formatted before writing it to the AIC, and the input data from the AIC must be formatted properly before it is used in calculations. The data word should first be left-shifted 16 bits to move the sign bit to the MSB of the 32-bit TMS320C30 integer word. The data should then be arithmetically right-shifted 18 bits to align the LSB of the AIC input data with the LSB of the 32-bit TMS320C30 data word. This procedure extracts the relevant 14 bits from the AIC input word and sign-extends the result.

Figure 5.3 shows a C30 program, which is supplied on the included disk, that implements an FIR filter using the techniques described above. The filter coefficients represent a low-pass filter designed by the window design method using a rectangular window. The passband is specified as 0 to 1.2 kHz with a sampling frequency of 8 kHz. The filter coefficients can be generated using the Matlab function `firbox` in Appendix C.

A couple of features of the `EVM30` debugger is worth pointing out before you start working with the FIR program. To view the array `hnsec` as floating-point numbers, use the command: `disp *( float *) hnsec`. (Notice for this to work, `hnsec` must be declared as a `.global` in the program.) It will bring up a new window with the floating-point array entries listed sequentially. To view the content of a register as a floating-point value, use the command `DISP Fx` where `x` is one of the eight the registers;

i.e., to view register R3 as a float, simply type DISP F3.

```
*****************************************************************************
*                                                                           *
*                       FIR filtering Program - FIR.ASM                     *
*                       -------------------------------                     *
*                                                                           *
*   This program implements an FIR filter using circular addressing         *
*   to manage buffer with input samples.                                    *
*   The program calls evm_init to initialize the EVM hardware board         *
*                                                                           *
*      Henrik Sorensen.                                                     *
*      Written in Sep. 14, 1990. Modified in Feb 1, 1995. Version 1.0       *
*                                                                           *
*  Do not distribute without permission from authors. Copyright 1995-96     *
*                                                                           *
*****************************************************************************

           .global start               ;entry point of the program
           .global wait_intr           ;wait for interrupt

* Symbols defined in "sysinit"
           .global evm_init            ;EVM initialization subroutine
           .global receive0            ;receive0 interrupt routine
           .global p0_addr             ;serial port 0 address
           .global stack_addr          ;system stack address

* Constants for assembler
N          .set    7                   ;specify filter length

* Define section for input data
           .global xnsec
           .sect   "inpsam"
xnsec:     .space  7

* Define section with filter coefficients
           .global hnsec
           .sect   "coeffs"
hnsec:     .float  2.770772250333945e-02
           .float  1.279134021017221e-01
           .float  2.176192777937789e-01
           .float  2.535191952023190e-01
           .float  2.176192777937789e-01
           .float  1.279134021017221e-01
           .float  2.770772250333945e-02

           .text
* Words needed for de-referencing
hn:        .word   hnsec
xn:        .word   xnsec

* Initialize system first
```

```
start:      ldi     @stack_addr, sp       ;load stack pointer
            call    evm_init              ;routine to initialize the EVM

* Initialize global variables
            ldi     @p0_addr,ar0          ;get port address
            ldi     @xn, ar1              ;ar1 points to x(n-(N-1))
            ldi     @hn, ar2              ;ar0 points to h(N-1)
            ldi     N, bk                 ;set block size

* Wait in idel loop for interrupts
wait_intr: idle                           ;wait for interrupts
            br      wait_intr

* Interrupt service routine

receive0:   ldi     *+ar0(12),r0          ;read a sample

* Convert from xxxxxxxxxxxxxx00 format to normal sign-extended 14-bit format
            lsh     16, r0                ;move 14 data bits to msb boundary
            ash     -18, r0               ;move 14 data bits to 14 lsb bits
            float   r0, r6                ;convert  to float
            stf     r6, *ar1++%           ;store the sample
            ldf     0.0, r0               ;initialize r0 to 0.0
            ldf     0.0, r2               ;initialize r2 to 0.0

            rpts    N-1                   ;for i=0  to N-1
            mpyf    *ar2++%, *ar1++%, r0  ;r0=a(N-1-i)*x(n-(N-1-i))
||          addf    r0, r2, r2            ;accumulate products

            addf    r0, r2                ;add last product to sum
* Convert to xxxxxxxxxxxxxx00 format
            fix     r2, r2                ;Convert to integer
            lsh     2, r2                 ;shift left two bits
* Write converted word to D/A
            sti     r2,*+ar0(8)           ;write output sample to D/A
            reti                          ;return from interrupt

            .end
```

Figure 5.3: TMS320C30 assembly program for FIR filtering

5.2.1 Experiment 5A: FIR Filter Implementations

5A.1 Create an `FIR.cmd` file for the FIR program shown in Fig. 5.3, compile, and run program. Hint: The program is using circular increments of both `AR1` and `AR2`, so care must be taken when creating `fir.cmd`. Measure magnitude frequency response using function generator and oscilloscope.

h[0]	=	h[24]	=	-2.617402054584e-02
h[1]	=	h[23]	=	-2.428903825652e-02
h[2]	=	h[22]	=	0.000000000000e+00
h[3]	=	h[21]	=	2.968660231353e-02
h[4]	=	h[20]	=	3.926103081877e-02
h[5]	=	h[19]	=	1.457906537991e-02
h[6]	=	h[18]	=	-3.235286863914e-02
h[7]	=	h[17]	=	-6.605038526494e-02
h[8]	=	h[16]	=	-4.852930295871e-02
h[9]	=	h[15]	=	3.401781921980e-02
h[10]	=	h[14]	=	1.570441232751e-01
h[11]	=	h[13]	=	2.671794208217e-01
		h[12]	=	3.112551076727e-01

Table 5.1: Filter coefficients in file `firbox25.asc` for length 25 FIR filter designed using the window method with a rectangular window

5A.2 Table 5.1 lists the filter coefficients for a length-25 low-pass filter designed using the same specifications as before (rectangular window design method with a passband from 0 to 1.2 kHz and a sampling frequency of 8 kHz.) The filter coefficients can be found in the file `firbox25.asc` on the supplied disk. They can also be generated using the Matlab function `firbox` in Appendix C. Modify the program in Fig. 5.3 to implement this new filter and measure frequency response. How has frequency response changed now that the filter is much longer?

5A.3 Table 5.2 lists the coefficients for a lowpass filter with the same specifications as in Table 5.1, but with the filter order increased to 63. The filter coefficients can be found in the file `firbox63.asc` on the supplied disk. They can also be generated using the Matlab function `firbox` in Appendix C. Implement this filter with your program. Compare the results with the results from 5A.1 for the actual cut-off frequency, transition width, stopband attenuation, and ripple.

5A.4 Table 5.3 and 5.4 list the filter coefficients of two band-pass filters designed using the window technique with a Hamming window and the Parks-McClellan algorithm, respectively. The specifications for the Parks-McClellan filter is:

h[0]	=	h[62]	=	-8.272666448182e-03
h[1]	=	h[61]	=	3.753950284126e-17
h[2]	=	h[60]	=	8.843195168746e-03
h[3]	=	h[59]	=	1.076707795690e-02
h[4]	=	h[58]	=	3.628007391632e-03
h[5]	=	h[57]	=	-7.166298608955e-03
h[6]	=	h[56]	=	-1.267971682557e-02
h[7]	=	h[55]	=	-7.763490159701e-03
h[8]	=	h[54]	=	4.258965198872e-03
h[9]	=	h[53]	=	1.370355376333e-02
h[10]	=	h[52]	=	1.221203142351e-02
h[11]	=	h[51]	=	0.000000000000e+00
h[12]	=	h[50]	=	-1.349750841545e-02
h[13]	=	h[49]	=	-1.674878793296e-02
h[14]	=	h[48]	=	-5.762129386709e-03
h[15]	=	h[47]	=	1.164523523955e-02
h[16]	=	h[46]	=	2.113286137595e-02
h[17]	=	h[45]	=	1.330884027377e-02
h[18]	=	h[44]	=	-7.535092274927e-03
h[19]	=	h[43]	=	-2.512318189944e-02
h[20]	=	h[42]	=	-2.331387817215e-02
h[21]	=	h[41]	=	0.000000000000e+00
h[22]	=	h[40]	=	2.849473998818e-02
h[23]	=	h[39]	=	3.768477284917e-02
h[24]	=	h[38]	=	1.399374279629e-02
h[25]	=	h[37]	=	-3.105396063880e-02
h[26]	=	h[36]	=	-6.339858412784e-02
h[27]	=	h[35]	=	-4.658094095821e-02
h[28]	=	h[34]	=	3.265206652469e-02
h[29]	=	h[33]	=	1.507390913967e-01
h[30]	=	h[32]	=	2.564526598936e-01
		h[31]	=	2.987587892160e-01

Table 5.2: Filter coefficients from file `firbox63.asc` for length 63 FIR filter designed using the window method with a rectangular window

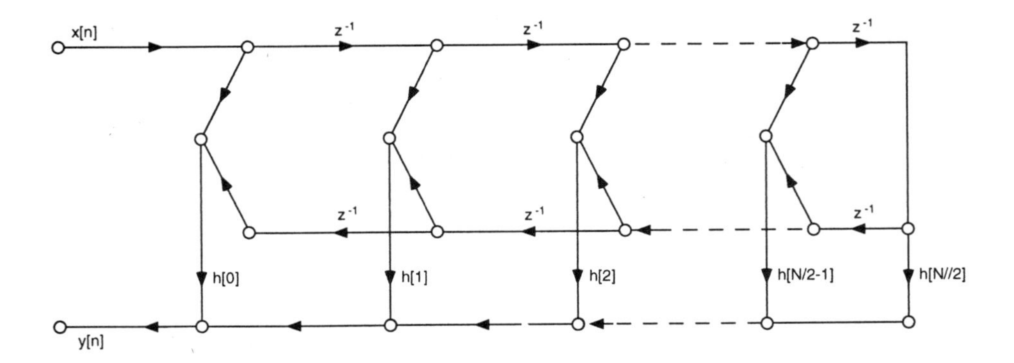

Figure 5.4: Direct form implementation of linear-phase FIR filter

```
Sampling frequency: 8 kHz
Lower Stopband: 0 - 0.6 kHz
Passband: 1 - 2 kHz
Upper Stopband: 2.4 - 4 kHz
Stopband Attenuation: 40 dB
```

and the window filter is chosen so it satisfies the same specifications. The filter coefficients can be found on the supplied disk in the files `firham63.asc` and `firrem42.asc`, respectively. They can also be generated using the two Matlab functions `firham.m` and `firrem.m` in Appendix C. Implement the two filters with your program using the circular memory approach. Compare the results.

5A.5 Modify the C30 program such that it does not use circular addressing. Compared to the original FIR C30 program, how many more instructions are needed to implement a length-100 filter without circular shift?

5A.6 Many FIR filters have linear phase that constrains their coefficients to be symmetrical. These filter can be implemented as shown in Fig. 5.4, which uses fewer multiplications than the normal Direct form in Fig. 5.1. Write a C30 program that uses the structure in Fig. 5.4 to implement the linear phase filter in Table 5.1. How many instructions does your program take to compute the length 25 filter? How does that compare with the program in Fig. 5.3?

5A.7 Modify the Matlab function `firbox` to accept several different window types and compare the frequency responses of the resulting filters.

5A.8 Design a highpass filter with a cut-off frequency of 2.5 kHz, a stopband attenuation of 40 dB, and a ripple of 0.5 dB using the Parks-

h[0]	=	h[62]	=	-2.411858027345e-04
h[1]	=	h[61]	=	8.760114839762e-04
h[2]	=	h[60]	=	1.679550861080e-03
h[3]	=	h[59]	=	4.093078086700e-18
h[4]	=	h[58]	=	-2.366098461639e-03
h[5]	=	h[57]	=	-1.691187474499e-03
h[6]	=	h[56]	=	6.072953993039e-04
h[7]	=	h[55]	=	0.000000000000e+00
h[8]	=	h[54]	=	-9.064564323019e-04
h[9]	=	h[53]	=	3.748318391300e-03
h[10]	=	h[52]	=	7.695533803821e-03
h[11]	=	h[51]	=	-1.912552993540e-17
h[12]	=	h[50]	=	-1.090130629140e-02
h[13]	=	h[49]	=	-7.528764577165e-03
h[14]	=	h[48]	=	2.585767159503e-03
h[15]	=	h[47]	=	0.000000000000e+00
h[16]	=	h[46]	=	-3.509707089758e-03
h[17]	=	h[45]	=	1.389541430922e-02
h[18]	=	h[44]	=	2.745817964940e-02
h[19]	=	h[43]	=	0.000000000000e+00
h[20]	=	h[42]	=	-3.677275313276e-02
h[21]	=	h[41]	=	-2.499512165319e-02
h[22]	=	h[40]	=	8.531489287950e-03
h[23]	=	h[39]	=	0.000000000000e+00
h[24]	=	h[38]	=	-1.186973998752e-02
h[25]	=	h[37]	=	4.879700629988e-02
h[26]	=	h[36]	=	1.026473297665e-01
h[27]	=	h[35]	=	0.000000000000e+00
h[28]	=	h[34]	=	-1.777434851091e-01
h[29]	=	h[33]	=	-1.580430285800e-01
h[30]	=	h[32]	=	9.323892374661e-02
		h[31]	=	2.506132399314e-01

Table 5.3: Filter coefficients from file `firham63.asc` for length 63 band-pass filter designed with Hamming window method

h[0]	=	h[41]	=	-1.470143798961e-02
h[1]	=	h[40]	=	-3.425062753415e-02
h[2]	=	h[39]	=	7.927907924993e-03
h[3]	=	h[38]	=	1.549192212063e-03
h[4]	=	h[37]	=	3.146292825687e-03
h[5]	=	h[36]	=	1.932446314514e-02
h[6]	=	h[35]	=	2.316862655573e-02
h[7]	=	h[34]	=	-9.065935162281e-03
h[8]	=	h[33]	=	-1.889518332511e-02
h[9]	=	h[32]	=	-4.117659850422e-03
h[10]	=	h[31]	=	-2.578989948655e-02
h[11]	=	h[30]	=	-3.815414399111e-02
h[12]	=	h[29]	=	2.459011407107e-02
h[13]	=	h[28]	=	6.054938235587e-02
h[14]	=	h[27]	=	1.077162095151e-02
h[15]	=	h[26]	=	1.736587971863e-02
h[16]	=	h[25]	=	6.877740462501e-02
h[17]	=	h[24]	=	-5.873628648392e-02
h[18]	=	h[23]	=	-2.455006273052e-01
h[19]	=	h[22]	=	-9.019135319685e-02
h[20]	=	h[21]	=	2.750257304823e-01

Table 5.4: Filter coefficients from file `firrem42.asc` for length 42 Parks-McClellan FIR filter designed with program `firrem` in Appendix C.10

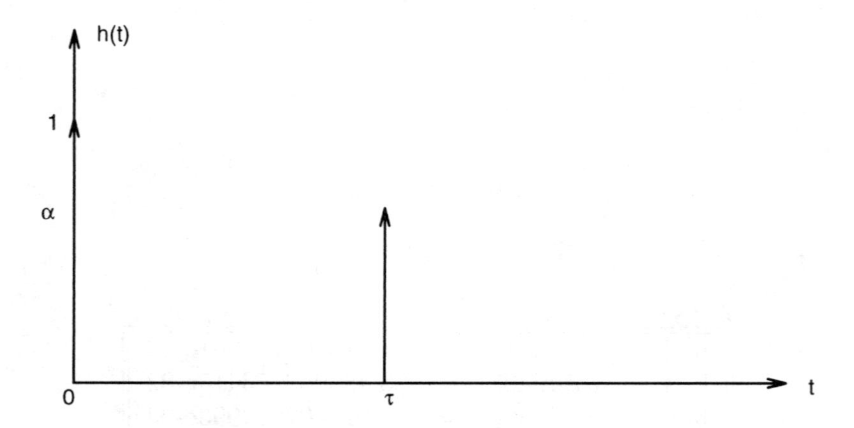

Figure 5.5: Impulse response for echo generator

McClellan technique. Measure the actual frequency response of your filter.

5A.9 Using the C interface in Appendix B, call your FIR filter routine from a C main program.

5A.10 Using the Matlab interface in Chapter 11, create a Matlab external function C30FIR, which should be invoked as: `y=C30FIR(x,h);`. It transfers the two sequences `x` and `h` to the TMS320C30 from Matlab, convolves them on the TMS320C30, and transfers the result back to Matlab.

5.2.2 Experiment 5B: Echo Generation

An echo can be thought of as a delayed, repeated version of the signal with some attenuation. Hence, the impulse response of an echo generator could look something like that shown in Fig. 5.5. (Notice the impulse at the origin.)

5B.1 Implement the echo generator with a delay τ of 1.0 sec. and an attenuation α of 0.5. Use a sampling frequency of 4 kHz; i.e., $A = 26$ and $B = 36$. To solve this problem, it is useful to refer to the memory map for the EVM shown in Fig. 2.7 before you write your program. Use the oscilloscope and function generator to verify your echo generator with a low frequency sinewave as input. Change the frequency slightly and explain what you see.

5B.2 Change your program to generate a pure delay (i.e., remove the impulse at the origin from the impulse response in Fig. 5.5). Hook up a microphone and speaker as input/output, respectively. Now slowly increase your delay τ to a point where you clearly can hear the time difference. What is this value of τ? Does this threshold value depend on the value of α? Notice to get enough memory for the delay buffer, you will have to locate it in SRAM.

5.2.3 Experiment 5C: Spectrum Reversal

Assume we are given an input signal with a spectrum as indicated in Fig. 5.6(a) and that the desired output spectrum is the "mirrored" version of that as shown in Fig. 5.6(b). The obvious solution is to convert the problem to the frequency domain, but that is not necessary.

5C.1 Find an alternate method for implementing the spectrum reversal in the time domain, which involves multiplying by some constants that can be implemented *mostly like* an FIR filter but with coefficients that "change." (Hint: Spectrum reversal is really a 180 degree phaseshift)

5C.2 Implement your solution.

5C.3 Use the function generator and oscilloscope to sweep the input frequency from DC to 4 kHz. Explain what you see.

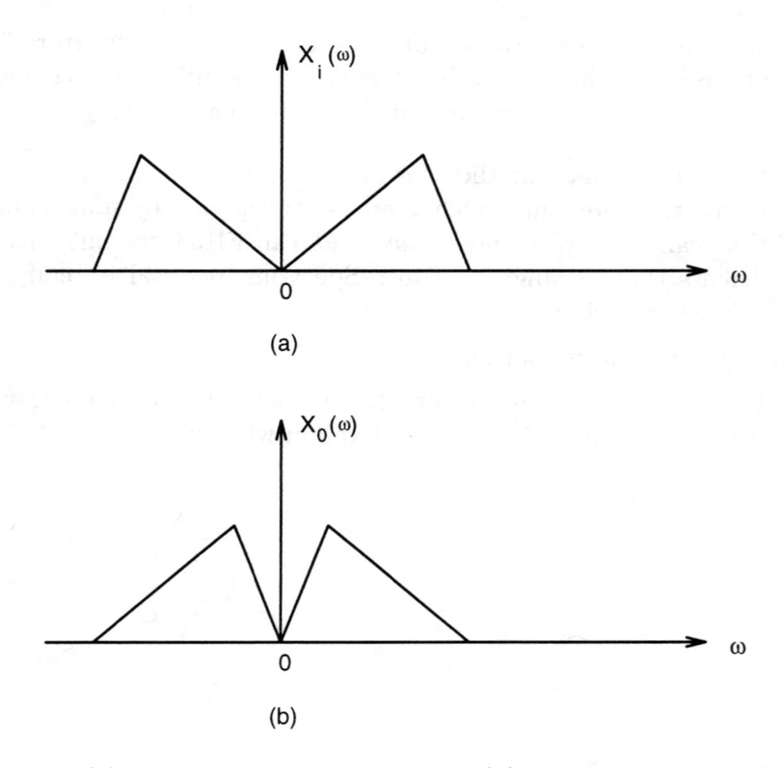

Figure 5.6: (a) Spectrum of input signal; (b) Desired output spectrum, which is the mirror of the input spectrum

Chapter 6

IIR Filter Implementations

6.1 IIR Filters

In this chapter, we discuss the implementation of Infinite Impulse Response (IIR) filters on the TMS320C30. Compared to FIR filters, IIR filters are often much more efficient in terms of attaining better magnitude response for a given filter order. This is because IIR filters incorporate feedback and thus are capable of realizing both poles and zeros of a system function, whereas FIR filters are only capable of realizing the zeros. This means that IIR filters can run faster since lower orders require fewer computational steps. They can, however, have stability problems and always have nonlinear phase characteristics, while FIR filters are always stable and can give exact linear phase.

The system function of an IIR filter is given by

$$\begin{aligned} H(z) &= \frac{Y(z)}{X(z)} \\ &= \frac{b_0 + b_1 z^{-1} + \ldots + b_N z^{-N}}{1 - a_1 z^{-1} - \ldots - a_N z^{-N}} \end{aligned} \tag{6.1}$$

where b_k and a_k are the coefficients of the filter, and the numerator and the denominator polynomials are assumed to be of the same order N. The difference equation representation of an IIR filter is expressed as:

$$\begin{aligned} y[n] = \; & b_0 x[n] + b_1 x[n-1] + \ldots + b_N x[n-N] + \\ & a_1 y[n-1] + a_2 y[n-2] + \ldots + a_N y[n-N] \end{aligned} \tag{6.2}$$

This representation implies that the output of an IIR filter is a function of the past outputs and the present and past inputs.

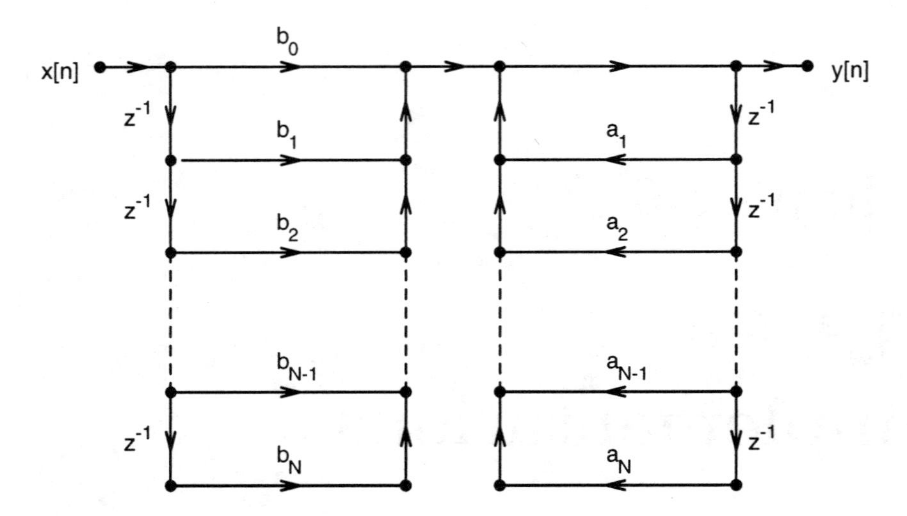

Figure 6.1: Direct Form 1 structure for an IIR filter

There exist several different structures that can be used to realize an IIR filter. In our experiments, we will use two different structures: the Direct Form and the Cascade Form. The Direct Form, as the name suggests, implements the difference equation directly. Since there are two parts to this filter, namely the FIR and the recursive (or equivalently, the numerator and denominator) parts, this implementation actually consist of two versions, Direct Form 1 and Direct Form 2. The structure of the Direct Form 1 is shown in Fig. 6.1. It implements each of the two parts separately with a cascade connection between them. The numerator, or FIR part, is a tapped delay line followed by the denominator, or recursive part, which is a feedback tapped delay line. Thus, there are two separate delay lines in this structure and therefore requires two separate sections of memory elements.

We can reduce this memory requirement by eliminating one of the delay lines. By interchanging the order in which the two parts are connected in the cascade, the two delay lines are beside each other, connected by a unity gain branch. Therefore, one of the delay lines can be removed, which leads to the Direct Form 2 structure, shown in Fig. 6.2.

As can be seen, the Direct Form 2 is more efficient in terms of the memory usage and speed. It should be noted that both of the Direct Forms are equivalent from the input-output point of view. Internally, however, they have different signals. The difference equation for the Direct Form 2 can be rewritten as

$$w[n] = x[n] + a_1 w[n-1] + a_2 w[n-2] + \ldots + a_N w[n-N] \tag{6.3}$$

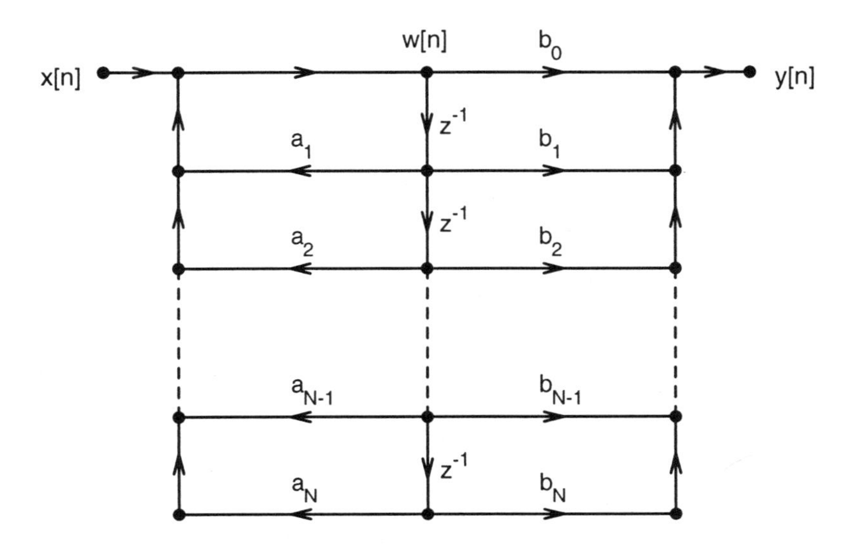

Figure 6.2: Direct Form 2 structure for an IIR filter

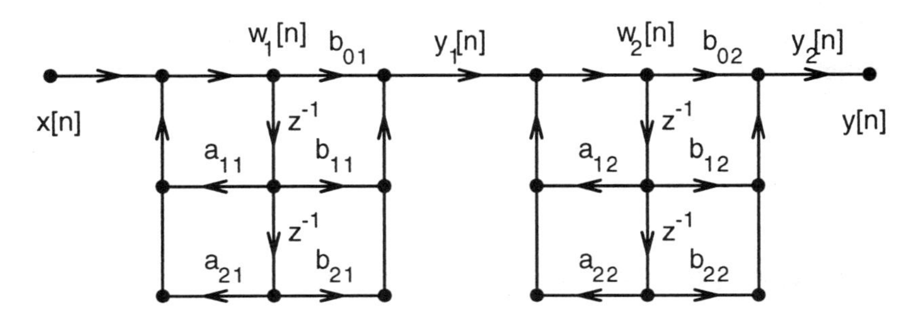

Figure 6.3: Cascade structure for a fourth-order IIR filter

$$y[n] = b_0 w[n] + b_1 w[n-1] + \ldots + b_N w[n-N] \tag{6.4}$$

In the Cascade Form, the system function $H(z)$ in Eqn. (6.1) is factored into smaller second-order sections, called biquads. The system function can then be represented as a product of these biquads,

$$H(z) = \prod_{k=1}^{N_s} \frac{b_{0k} + b_{1k} z^{-1} + b_{2k} z^{-2}}{1 - a_{1k} z^{-1} - a_{2k} z^{-2}} \tag{6.5}$$

where N_s is the integer part of $(N+1)/2$.

With each biquad implemented in a Direct Form, the entire system becomes a cascade of biquad sections. Figure 6.3 shows a cascade structure for a fourth order system using two Direct Form 2 second-order sections.

The difference equations representing a general cascade of Direct Form 2 second-order sections are of the form

$$y[0,n] = x[n] \tag{6.6}$$

$$w[k,n] = a_{1k}w[k,n-1] + a_{2k}w[k,n-2] + y[k-1,n] \quad k = 1,2,\ldots,N_s \tag{6.7}$$

$$y[k,n] = b_{0k}w[k,n] + b_{1k}w[k,n-1] + b_{2k}w[k,n-2] \quad k = 1,2,\ldots,N_s \tag{6.8}$$

$$y[n] = y[N_s,n] \tag{6.9}$$

As indicated before, the IIR filters can have stability problems that are more likely to occur when the order is large. The Cascade Form solves this problem by breaking one large filter up into several smaller second-order sections and hence reducing the filter sensitivity to coefficient quantization [9]. The Cascade structure generally also introduces less quantization noise than the Direct structures because it limits the dynamic range of the signals by appropriately grouping the zeros and poles into each second-order section [9].

The basic IIR filter design technique is the transformation of an analog filter into a digital filter using a complex-valued mapping. There are several such transformations and the most popular include the Impulse Invariant Transformation and the Bilinear Transformation. Like the case of FIR filters, the design techniques of IIR filters are well described in many digital signal processing texts [8, 9, 11], and, hence, will not be repeated here.

6.2 Implementation of IIR Filters

Coefficient quantization can have a severe effect on the performance of the IIR filter. This makes the implementation of IIR filters more complicated, especially if using fixed-point arithmetic. Even in the case of the cascade structure, one must carefully choose a set of scale factors and perform scaling within each biquad to avoid an overall overflow and instability problem. As in the case of FIR filters, the floating-point ability of the TMS320C30 simplifies the implementation of the IIR filter.

It is obvious that the implementation of an IIR filter is similar to that of an FIR filter, since they are described by similar difference equations. Therefore, all of the features of the TMS320C30 that facilitate the implementation of an FIR filter, such as the parallel multiply/add operation, the `RPTS` or `RPTB` instruction, and the circular addressing, also apply to the IIR filters.

For the structure of the Direct Form 1, the two convolution sums can be implemented as two separate "FIR filters" over the past inputs and outputs respectively. The difference between these two sums become the next output value. In DirectForm 1, this requires two circular buffers to hold the input $x[n]$ and output $y[n]$. However, in the Direct Form 2, the input and output delay lines are incorporated into one delay line—the intermediate signal $w[n]$. Thus, only one circular buffer is needed. Nonetheless, two convolution sums are computed from $w[n]$, which must be calculated first. Figure 6.4 shows a Direct Form 2 implementation of the filter shown in Eqn. (6.10) below. Notice the scaling by $G = 0.0074$ right before the result is transferred to the D/A converter. Whereas the DSP is a floating-point device, the D/A converter is a fixed-point 14-bit device. We, thus, need to ensure that the signal has amplitudes that do not exceed the range -2^{13} to $2^{13} - 1$ to prevent overflow, but that is not much smaller so as to limit the effect of the quantization noise. The filter in Eqn. (6.10) has a gain of 135.59, so to ensure unity gain we must scale with $\frac{1}{135.59} = 0.0074$.

```
*****************************************************************************
*                                                                           *
*                        IIR filtering Program - IIR.ASM                    *
*                        ------------------------------                     *
*                                                                           *
*   This program implements a Direct Form 2 IIR filter using circular       *
*   addressing to manage buffer with input samples.                         *
*   The program calls evm_init to initialize the EVM hardware board         *
*                                                                           *
*      Henrik Sorensen.                                                     *
*      Written in Sep. 14, 1990. Modified in Feb 1, 1995. Version 1.0       *
*                                                                           *
*  Do not distribute without permission from authors. Copyright 1995-96     *
*                                                                           *
*****************************************************************************

           .global start                  ;entry point of the program
           .global wait_intr              ;wait for interrupt

* Symbols defined in "sysinit"
           .global evm_init               ;EVM initialization subroutine
           .global receive0               ;receive0 interrupt routine
           .global p0_addr                ;serial port 0 address
           .global stack_addr             ;system stack address

* Constants for assembler
N          .set    4                      ;specify filter length

* Define section for state variables and filter coefficients
           .sect   "filt"
wstate:    .float  0.0                    ;state variables for filter
           .float  0.0
           .float  0.0
           .float  0.0
```

```
bcoef:    .float  1.0               ;b[4]
          .float  4.0               ;b[3]
          .float  6.0               ;b[2]
          .float  4.0               ;b[1]
acoef:    .float  -0.5166           ;a[4]
          .float  1.9198            ;a[3]
          .float  -3.1119           ;a[2]
          .float  2.5907            ;a[1]

          .text
* Words needed for de-referencing
a:        .word   acoef
b:        .word   bcoef
w:        .word   wstate
G:        .float  0.0074             ;1/gain of filter

* Initialize system first
start:    ldi     @stack_addr, sp     ;load stack pointer
          call    evm_init            ;routine to initialize the EVM

* Initialize global variables
          ldi     @p0_addr,ar3        ;get port address
          ldi     @w, ar0             ;circular buffer address
          ldi     N, bk               ;circular buffer size

* Wait in idle loop for interrupts
wait_intr: idle                       ;wait for interrupts
          br      wait_intr

* Interrupt service routine

receive0: ldi     *+ar3(12),r2        ;read a sample

* Convert from xxxxxxxxxxxxxx00 format to normal sign-extended 14-bit format
          lsh     16, r2              ;move 14 data bits to msb boundary
          ash     -18, r2             ;move 14 data bits to 14 lsb bits
          float   r2, r3              ;convert  to float

          ldf     0.0, r0             ;initialize r0 to 0.0
          ldi     @a, ar1             ;pointer to filter coefficients

          rpts    N-1                 ;for i=0  to N-1
          mpyf3   *ar0++%, *ar1++, r0 ;r0=a[i]*w[N-i]
||        addf3   r0, r3, r3          ;accomulate products

          addf    r0, r3              ;add last product to sum
          ldf     r3, r4              ;keep copy of w[n] in r4

          ldf     0.0, r1
          ldi     @b, ar2
          rpts    N-1                 ;for i=0  to N-1
```

```
          mpyf3     *ar0++%, *ar2++, r1   ;r0=b[i]*w[N-i]
||        addf3     r1, r3, r3            ;accomulate products

          addf      r1, r3                ;add last product to sum

          stf       r4, *ar0++%           ;store w[n]

          mpyf      @G, r3                ;multiply with G
* Convert to xxxxxxxxxxxxxx00 format
          fix       r3, r5                ;Convert to integer
          lsh       2, r5                 ;shift left two bits
* Write converted word to D/A
          sti       r5,*+ar3(8)           ;write output sample to D/A
          reti                            ;return from interrupt

          .end
```

Figure 6.4: TMS320C30 assembly program for IIR filtering

Often, IIR filters are constructed using the cascade form of second-order sections, the biquads. In this structure, each biquad is usually implemented in the Direct Form 2. For each biquad, a length-3 circular buffer is used to store $w[k, n]$. Thus, N_s length-3 circular buffers are required for a N_s biquad system. As in the case of FIR filters, some care should be taken in allocating the buffers in memory. The beginning address of each $w[k, n]$ buffer must be a multiple of 4; i.e., its last two bits must be zero. The block-size register BK should be initialized to 3. Figure 6.5 shows an example of the memory organization for N_s biquads.

6.2.1 Experiment 6A: IIR Filter Implementations

6A.1 Figure 6.4 shows a program that implements a 4th-order Chebychev lowpass IIR filter (designed using bilinear transformation method), with a passband from 0 to 1.2 kHz and a sampling rate of 8 kHz, is given by

$$H(z) = \frac{z^4 + 4z^3 + 6z^2 + 4z + 1}{z^4 - 2.5907z^3 + 3.1119z^2 - 1.9198z + 0.5166} \tag{6.10}$$

Create an `IIR.cmd` file for the program, compile, and run it. Measure the magnitude frequency response using a function generator and oscilloscope.

6A.2 The transfer function in experiment 6A.1 can be written as a prod-

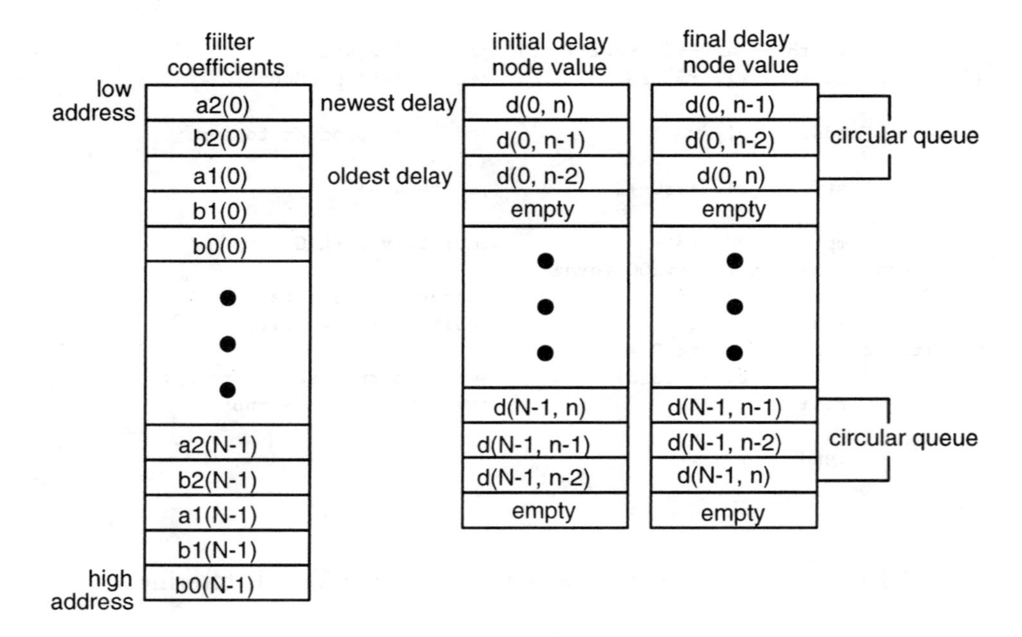

Figure 6.5: Data Memory Arrangement for N_s biquads.

uct of two second-order sections:

$$H(z) = \frac{z^2 + 2z + 1}{z^2 - 1.1475z + 0.8430} \times \frac{z^2 + 2z + 1}{z^2 - 1.4432z + 0.6128} \tag{6.11}$$

Write a C30 program to implement the filter in the cascade form. How do you handle the filter gain in this case?

6A.3 Table 6.1 and 6.2 give, respectively, the coefficients for a Direct Form and Cascade Form structure for a Butterworth lowpass filter of order 17 with the following specifications:

```
Sampling frequency: 8 kHz
Passband: 0 -1 kHz
Stopband: 1.4 - 4 kHz
Passband Ripple: 1 DB
Stopband Attenuation: 50 DB
```

The direct and cascade form of the filter are obtained using the `dfbutter` and `casbutter` Matlab routines in Appendix C, respectively, which uses a bilinear transformation method from an analog Butterworth prototype filter. The filter coefficients can also be found in the files `dfbut17.asc` and `casbut17.asc` on the supplied disk. Implement the two different structures of the filter and compare the resulting frequency responses.

6A.4 Table 6.3 gives the cascade form filter coefficients for a filter with the same specifications as in 6A.3. The filter is also designed using the bilinear transformation method, but from an analog Chebychev filter, which results in a filter order of 8. The filter coefficients are in file `caschb17.asc` on the supplied disk and can be generated using the Matlab function `cascheby.m` in Appendix C. Implement the filter and compare it with the one in 6A.3.

6A.5 Write a Matlab script that will generate the "Direct Form" Butterworth filter coefficients in a format that can be directly inserted into the assembler source file. Use your script to vary the filter order from N=2 to 20 and plot the cut-off at 1.0 kHz versus filter order. Comment on your findings.

6A.6 Implement a length 9 high-pass Butterworth filter using both the Direct Form and Cascade Form. This can be done easily by modifying the `dfbutter` and `casbutter` Matlab programs in Appendix C.

6A.7 Change the coefficients of the filter in 6A.1 to make the filter unstable. Run the filter and record your observations.

Numerator coefficients		Denominator coefficients	
b(0) =	3.7894798410e-09	a(0) =	1.0000000000e+00
b(1) =	b(0) * 17 = 6.4421154633e-08	a(1) =	8.4923153496e+00
b(2) =	b(0) * 136 = 5.1536931522e-07	a(2) =	-3.5212366439e+01
b(3) =	b(0) * 680 = 2.5768460148e-06	a(3) =	9.4114421456e+01
b(4) =	b(0) * 2380 = 9.0189626576e-06	a(4) =	-1.8085295172e+02
b(5) =	b(0) * 6188 = 2.3449299988e-05	a(5) =	2.6433603033e+02
b(6) =	b(0) * 12376 = 4.6898603955e-05	a(6) =	-3.0355365317e+02
b(7) =	b(0) * 19448 = 7.3697802009e-05	a(7) =	2.7919517209e+02
b(8) =	b(0) * 24310 = 9.2122256092e-05	a(8) =	-2.0787780554e+02
b(9) =	b(0) * 24310 = 9.2122253719e-05	a(9) =	1.2583609689e+02
b(10) =	b(0) * 19448 = 7.3697804254e-05	a(10) =	-6.1855704239e+01
b(11) =	b(0) * 12376 = 4.6898602168e-05	a(11) =	2.4524821311e+01
b(12) =	b(0) * 6188 = 2.3449301260e-05	a(12) =	-7.7399267747e+00
b(13) =	b(0) * 2380 = 9.0189619779e-06	a(13) =	1.9020161343e+00
b(14) =	b(0) * 680 = 2.5768462881e-06	a(14) =	-3.5121559851e-01
b(15) =	b(0) * 136 = 5.1536925644e-07	a(15) =	4.5890998120e-02
b(16) =	b(0) * 17 = 6.4421157154e-08	a(16) =	-3.7861999061e-03
b(17) =	b(0) * 1 = 3.7894798286e-09	a(17) =	1.4843369588e-04

Table 6.1: Coefficients for length 17 IIR filter in Direct Form. Coefficients are generated using bilinear transformation of Butterworth prototype using `dfbutter` program in Appendix C.6 and can be found in file `DFBUT17.ASC`.

Denominator coefficients		
a_{0k}	a_{1k}	a_{2k}
1.000000000000e+00	1.327596415868e+00	-8.775048566785e-01
1.000000000000e+00	1.184920779368e+00	-6.757310365197e-01
1.000000000000e+00	1.075296614113e+00	-5.206990552530e-01
1.000000000000e+00	9.916462732481e-01	-4.023996087043e-01
1.000000000000e+00	9.288403864227e-01	-3.135786717589e-01
1.000000000000e+00	8.832242036269e-01	-2.490676473853e-01
1.000000000000e+00	8.522657813874e-01	-2.052858267846e-01
1.000000000000e+00	8.343113332208e-01	-1.798944026825e-01
1.000000000000e+00	4.142135623731e-01	0.0
Numerator coefficients		
b_{0k}	b_{1k}	b_{2k}
1.374771102026e-01	2.749542204052e-01	1.374771102026e-01
1.227025642880e-01	2.454051285759e-01	1.227025642880e-01
1.113506102849e-01	2.227012205698e-01	1.113506102849e-01
1.026883338640e-01	2.053766677281e-01	1.026883338640e-01
9.618457133404e-02	1.923691426681e-01	9.618457133404e-02
9.146086093961e-02	1.829217218792e-01	9.146086093961e-02
8.825501134929e-02	1.765100226986e-01	8.825501134929e-02
8.639576736541e-02	1.727915347308e-01	8.639576736541e-02
2.928932188135e-01	2.928932188135e-01	0.0

Table 6.2: Coefficientsfor Cascade Form of length 17 Butterworth filter. Coefficients are generated using program `casbutter` in Appendix C.3 and can be found in file `CASBUT17.ASC`.

Denominator coefficients		
a_{0k}	a_{1k}	a_{2k}
1.000000000000e+00	1.382883636792e+00	-9.516463638369e-01
1.000000000000e+00	1.451624547971e+00	-8.631176328394e-01
1.000000000000e+00	1.592975535779e+00	-7.908702910505e-01
1.000000000000e+00	1.706506715689e+00	-7.481523422986e-01
Numerator coefficients		
b_{0k}	b_{1k}	b_{2k}
1.430287249570e-01	2.860574499140e-01	1.430287249570e-01
1.421799616717e-01	2.843599233433e-01	1.421799616717e-01
1.451440394812e-01	2.902880789623e-01	1.451440394812e-01
1.481814469001e-01	2.963628938003e-01	1.481814469001e-01

Table 6.3: Coefficients for length 8 Chebychev filter that satisfies the same specifications as the length 17 Butterworth filter in Table 6.1 and 6.2. Coefficients are generated using program `cascheby` in Appendix C.4 and can be found in file `CASCHB17.ASC`.

Chapter 7

Fast Fourier Transform

7.1 The Fast Fourier Transform

The discrete Fourier transform (DFT) is an important tool in many digital signal processing systems. It converts information from a time domain to a frequency domain representation. Methods that are computationally efficient for computing the DFT are known as the Fast Fourier Transform (FFT) algorithms [8]. In this chapter, we discuss implementation of FFTs on the TMS320C30.

The DFT of a finite-length sequence of length N is defined as

$$X[k] = \sum_{n=0}^{N-1} x[n] W_N^{nk} \qquad k = 0, 1, \ldots, N-1 \tag{7.1}$$

where W_N is the complex-valued phase factor $e^{-j2\pi/N}$. It is clear from Eqn. (7.1) that a direct calculation of the DFT summation requires N complex multiplications and $N-1$ complex additions for each of the N output samples. Consequently, to compute all N values of the DFT requires N^2 complex multiplications and $N^2 - N$ complex additions (This is equal to $4N$ real multiplications and $4N-2$ real additions). This direct computation of the DFT from Eqn. (7.1) is inefficient because it does not exploit the symmetry and periodicity of the phase factors W_N. Many methods that efficiently compute the DFT have been developed and they are collectively called FFT algorithms. In our experiments, we will focus on the radix-2 decimation-in-time (DIT) FFT algorithm. A Matlab program that explicitly implements the DIT FFT is included in Appendix C.

The decimation-in-time FFT divides the length N (where N is a power of 2) input sequence into two groups, one of the even indexed samples and the other of the odd indexed samples. Two length $N/2$ DFTs are then performed on these subsequences, and their outputs are combined to form the length N DFT. The methodology is illustrated by the following equations. First the input sequence in Eqn. (7.1) is divided into the even and the odd subsequences:

$$\begin{aligned} X[k] &= \sum_{n=0}^{\frac{N}{2}-1} x[2n]W_N^{2nk} + \sum_{n=0}^{\frac{N}{2}-1} x[2n+1]W_N^{(2n+1)k} \\ &= \sum_{n=0}^{\frac{N}{2}-1} x[2n]W_N^{2nk} + W_N^k \sum_{n=0}^{\frac{N}{2}-1} x[2n+1]W_N^{2nk} \qquad (7.2) \end{aligned}$$

By the substitutions

$$W_N^{2nk} = (e^{-j2\pi/N})^{2nk} = W_{\frac{N}{2}}^{nk} \qquad (7.3)$$

$$x_1[n] = x[2n] \qquad (7.4)$$

$$x_2[n] = x[2n+1] \qquad (7.5)$$

this equation becomes

$$\begin{aligned} X[k] &= \sum_{n=0}^{\frac{N}{2}-1} x_1[n]W_{\frac{N}{2}}^{nk} + W_N^k \sum_{n=0}^{\frac{N}{2}-1} x_2[n]W_{\frac{N}{2}}^{nk} \\ &= X_1[k] + W_N^k X_2[k] \qquad k = 0, 1, \ldots, N-1 \qquad (7.6) \end{aligned}$$

In Eqn. (7.6), $X_1[k]$ and $X_2[k]$ are two $N/2$ point DFTs and they can each be further divided into the even and odd sub-subsequences to form four $N/4$ point DFTs. In this way, by recursive decimation, the large DFT is broken into a combination of smaller DFTs. This is continued until only two-point DFTs, which are called radix-2 "butterflies," remain. The butterfly is the core calculation unit of the FFT. The W_N terms appear as coefficients (called twiddle factors) in the FFT calculation. The entire FFT is performed by computing butterflies in a "butterfly – group – stage" fashion in a three-loop algorithm. A complete 8-point DIT FFT is illustrated in Fig. 7.1.

Each pair of elements in Fig. 7.1, which are combined, represent a butterfly, and the entire FFT is made up of butterflies organized in different groups and stages. The first stage consists of four groups with one butterfly each. The second stage has two groups of two butterflies each, and the last stage has one group of four butterflies. Each butterfly has one multiplicative coefficient W_N^k and has two input points. Every stage contains $N/2$ (four) butterflies. For the more general case where N is a power

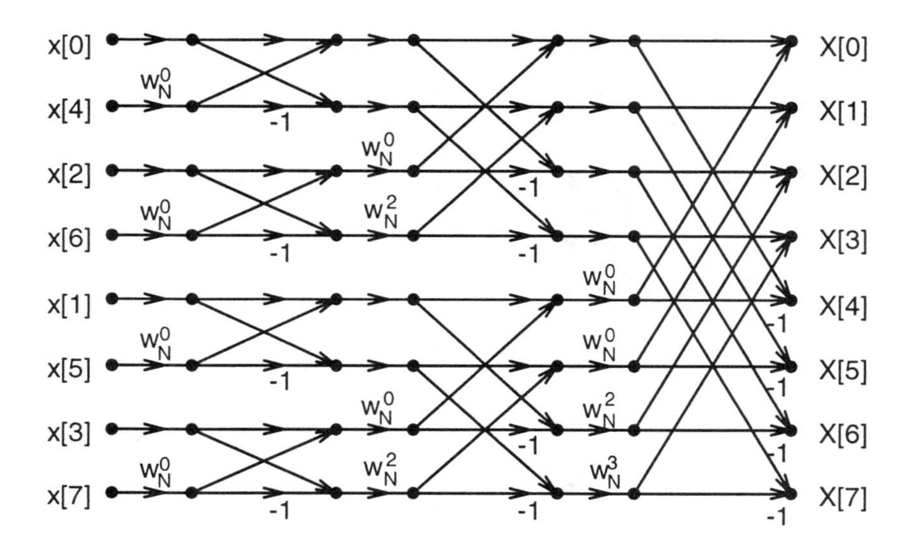

Figure 7.1: Eight point DIT FFT

of 2, the FFT will consist of $\log_2(N)$ stages and $\frac{N}{2}\log_2(N)$ butterflies. The computation of each butterfly requires one complex multiplication and two complex additions. Thus, the entire FFT is computed in $\frac{N}{2} * \log_2(N)$ complex multiplications and $N * \log_2(N)$ complex additions. Compared to the direct computation of the DFT, the number of the multiplications and additions required is greatly reduced.

Notice that whereas the output sequence is sequentially ordered, the input sequence is not. This is a consequence of repeatedly dividing the input sequence into subsequences of the even and the odd samples. Therefore, the input sequence must be scrambled before the butterfly calculations can begin. This is accomplished by a process called bit-reversal. Bit-reversal operates on the binary number that represents the index (position) of a sample within the array. The bit-reversed position has the index, which is the transpose of the binary number representing the original index; for example, the transpose of the 3-bit binary number 001 is 100.

The DIT FFT can appear in alternate, equivalent forms, one of which is shown in Fig. 7.2, where the input is in normal order but the output values are in a bit-reversed order.

The fundamental computational structure needed to compute the FFT is the generalized butterfly flow graph shown in Fig. 7.3. The variables x and y represent the real and imaginary parts, respectively, of a sample. The twiddle factor is also divided into its real and imaginary parts as $W_N^\alpha = e^{-j2\pi\alpha/N} = \cos(2\pi\alpha/N) - j\sin(2\pi\alpha/N)$ (represented as $C + j(-S)$). The

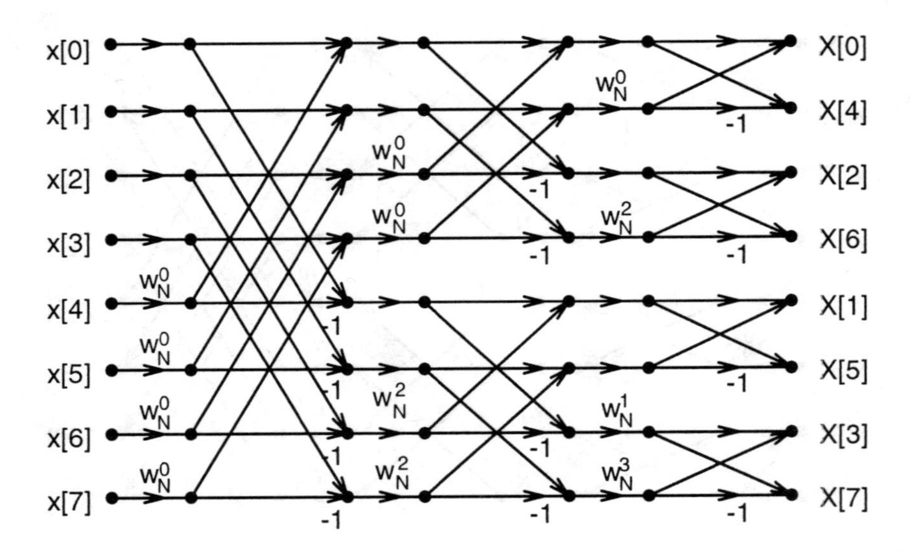

Figure 7.2: 8-point DIT FFT with input in normal order and output in bit-reversed order

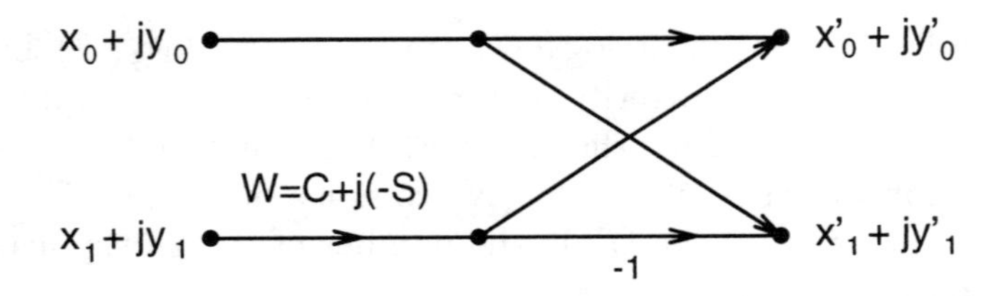

Figure 7.3: Radix-2 DIT FFT butterfly

lower input $(x_1 + jy_1)$ is multiplied by the twiddle factor $C + j(-S)$. The result of this multiplication is added to the upper input (x_0+jy_0) to produce the upper output $(x'_0+jy'_0)$ and subtracted from the upper input to produce the lower output $(x'_1 + jy'_1)$. Equations (7.7) through (7.12) yield the real and imaginary parts of the butterfly output.

$$temp_1 = C * x_1 - (-S) * y_1 \tag{7.7}$$

$$temp_2 = C * y_1 + (-S) * x_1 \tag{7.8}$$

$$x'_0 = x_0 + temp_1 \tag{7.9}$$

$$y'_0 = y_0 + temp_2 \tag{7.10}$$

$$x'_1 = x_0 - temp_1 \tag{7.11}$$

$$y'_1 = y_0 - temp_2 \tag{7.12}$$

The butterfly produces two complex outputs that become butterfly inputs in the next stage of the FFT. Since each stage has the same number of butterflies, the number of butterfly inputs and outputs remains constant from one stage to the next. This makes an "in-place" computation possible, which writes each butterfly output over the corresponding input (x_0' over x_0, etc.). In the in-place implementation, the FFT results end up in the same memory locations as the original inputs.

7.2 Implementation of FFTs

The calculation of the FFT is usually carried out in three nested loops, i.e., the stage, the group, and the butterfly loops. Among these three loops, the butterfly computation is the one repeated most frequently and can be implemented efficiently using the block-repeat (`RPTB`) instruction. All butterflies in one group can thus be computed within a single `RPTB` loop without any looping overhead. The two remaining loops, however, must be implemented in a brute force manner as the `RPTB` loop instruction can not be nested (Why not?).

The core part of an FFT algorithm is the computation of the butterfly. As given by Eqns. (7.7) through (7.12), the computation of each butterfly requires four real multiplications and six real additions. We would like to use four parallel multiply/add instructions to implement eight of these operations. The easiest way to achieve this is to compute the coefficient multiplications from each butterfly in parallel with the additions from the previous butterfly. Since there are $\frac{N}{2}\log_2(N)$ butterflies in a length N FFT, each operation removed from the butterfly calculation shaves $\frac{N}{2}\log_2(N)$ cycles of the time to compute the entire FFT, which is a significant number. For example, if $N = 1024$, we can save 0.3 mS for each instruction removed.

A hardware feature of the TMS320C30 that can increase the speed of an FFT, by up to 30 %, is the bit-reversed addressing mode. The DIT-FFT requires that the input data must be scrambled in a bit-reversed pattern before the butterfly calculations can start. This process is usually done by swapping data elements between various memory locations and, hence, requires extra instructions. However, the bit-reversed addressing feature, which is a special form of the indirect addressing mode, allows the bit-reversal to be done in conjunction with the read-in of the data without any extra cycles. In the bit-reversed addressing mode, one auxiliary register is used to point to the physical location of the data value in the sample array. This auxiliary register is incremented in a bit-reversed manner each time the register is accessed. For example, if $N = 8$, where N is the FFT length, and `base_adr` is the address of the first element in the input then the addresses will be: base_adr, base_adr+16, base_adr+8, base_adr+24, base_adr+4, base_adr+20, base_adr+12, base_adr+28. The index register IR0 must be set to one-half of the FFT length, i.e., IR0 contains the value $N/2$.

A TMS320C30 program that implements a decimation-in-time FFT according to the concepts above is shown in Fig. 7.4, which closely mimics the Matlab program shown in Appendix C.7. From these two programs, it should be evident that the FFT is computed in three nested loops indicated by the three labels: stage_loop, group_loop, and butterfly_loop in the TMS320C30 program. The twiddle-factor values are read from the address store at the SINE label. This table can be found in the program twf512.asm, which is on the included disk or can be found in Appendix D.1.

Since each butterfly adds its inputs, which are the output data from the previous stage, the computed values can grow by a factor of two in each stage of the FFT calculation. To ensure that no overflow occurs, an appropriate scaling arrangement is required if the FFT is implemented in fixed-point arithmetic. This troublesome task is eliminated on the TMS320C30 with its floating-point capability.

```
*****************************************************************************
*                                                                           *
*                        FFT program - FFT.ASM                              *
*                        ---------------------                              *
*                                                                           *
*  This program implements a decimation-in-time radix-2 FFT                 *
*  This program can not run as a stand alone, but needs fft_main.asm        *
*  move.asm, twf512.asm, evm_init.asm.                                      *
*                                                                           *
*     Henrik Sorensen.                                                      *
*     Written in Aug. 12, 1992. Modified in Feb 12, 1995. Version 1.0       *
*                                                                           *
* Do not distribute without permission from authors. Copyright 1995-96      *
*                                                                           *
*****************************************************************************

           .global fft                   ;Entry_point to fft subroutine
           .global N                     ;Length of FFT from fft_main.asm
           .global M                     ;log2(N) from fft_main.asm
           .global SINE                  ;Address of twiddle_factor table.
                                         ;Defined in twf512.asm
           .global fft_buf               ;Address of input/output buffer.
                                         ;Defined in fft_main.asm
           .text

fftsiz     .word   N                     ;length of FFT
logfft     .word   M                     ;log2(N)
```

```
sintab      .word   SINE                ;address of twiddle-factor table

fft:        ldi     @fftsiz,rc          ;Start of FFT
            subi    1,rc
            ldi     @fftsiz, ir0
            ldi     2,ir1
            ldi     @fft_buf, ar0
            ldi     ar0,ar1

            rptb    bitrv               ;in-place bit reversal
            cmpi    ar1,ar0
            bhs     no_exch
            ldf     *+ar0(1),r0
||          ldf     *+ar1(1),r1
            stf     r1,*+ar0(1)
||          stf     r0,*+ar1(1)
            ldf     *ar0,r0
||          ldf     *ar1,r1
            stf     r1,*ar0
||          stf     r0,*ar1
no_exch:    nop     *ar0++(ir1)
bitrv:      nop     *ar1++(ir0)B

            ldi     @fftsiz,ir1
            lsh     -2,ir1
            ldi     0,ar6               ;stage number
            ldi     4,ir0
            ldi     2,r7
            ldi     1, r6               ;group number
            ldi     @fftsiz,ar7
            negi    @logfft, ar3
            addi    9, ar3              ;for using a generic twiddle
            lsh     ar3, ir1            ;factor table up to 512 point

stage_loop:cmpi     @logfft,ar6
            bzd     end

            nop     *++ar6(1)           ;stage number counter
            lsh     -1,ar7
            lsh     ar3, ar7, r5
            ldi     1, ar5              ;group number counter
            ldi     0,ar1
            ldi     @sintab,ar4         ;ar4 points to sintab

group_loop:ldi      @fft_buf, ar0       ;ar0 points to X(0)
            addi    ar1, ar0            ;ar0 points to X(I)
            addi    r7,ar0,ar2          ;ar2 points to X(L)
            ldi     ar7,rc
            subi    1,rc
            ldf     *ar4,r4             ;r4=sin

            rptb    butterfly_loop      ;ar0 points to X(I)
```

```
          mpyf     *+ar2,r4,r0          ;Y(L)*S; ar2 points to X(L)
          mpyf     *ar2,*+ar4(ir1),r3   ;X(L)*C;
          mpyf     *+ar2,*+ar4(ir1),r0  ;Y(L)*C
||        addf     r0,r3                ;X(L)*C+Y(L)*S
          mpyf     *ar2,r4,r1           ;X(L)*S
||        addf     *ar0,r3,r2           ;X(I) + X(L) -> X(I)
          subf     r1,r0                ;Y(L)
          addf     *+ar0,r0,r1          ;Y(I) + Y(L) -> Y(I)
          subf     r0,*+ar0,r1          ;Y(I) - Y(L) -> Y(L)
||        stf      r1,*+ar0
          subf     r3,*ar0,r3           ;X(I) - X(L) -> X(L)
||        stf      r2,*ar0++(ir0)       ;ar0 -> X(I) of next group
butterfly_loop:
          stf      r3,*ar2++(ir0)       ;with the same TWF
||        stf      r1,*+ar2

          cmpi     r6,ar5
          bned     group_loop           ;branch to next group
          nop      *ar5++
          addi     r5, ar4
          addi     2,ar1

          brd      stage_loop           ;branch to next stage
          lsh      1, r6
          lsh      1, r7
          lsh      1,ir0

end:      rets
          .end
```

Figure 7.4: TMS320C30 assembly program for computing decimation-in-time FFT

7.2.1 Experiment 7A: Fast Fourier Transform Algorithms

The program in Fig. 7.4 only performs the FFT itself, it does not acquire data from the A/D converter and does not *do anything* with the resulting FFT coefficients.

7A.1 Compile the program in Fig. 7.4, and, using the debugger, verify that the program works for $N = 16$ by stepping through the program and comparing the data with the Matlab program in Appendix C.7. Use the twiddle-factors from the `twf512.asm` file in Appendix D.1. Why can these twiddle-factors be used for $N = 16$, even though they were generated for the $N = 512$ case? Also, why is there only one table, and not two (containing real and imaginary part of twiddle-factors)?

7A.2 Determine the number of cycles required to compute a length 16 and 128 transform. Next time the length-16 and length-128 FFT using an oscilloscope. To do this, let the FFT run in an infinite loop and transfer the value 128 to the output port at the beginning of each FFT and transfer −128 to the port at the end of each FFT. The time between the two pulses on the output now determines the time it takes to compute length-128 FFT. How does the time compare with the number of cycles determined earlier?

7A.3 The radix-2 DIT FFT implementation in Fig. 7.4 is a very compact form of the FFT in terms of the size of the program code. However, it is not the most efficient in terms of the execution speed because it does not exploit the characteristics of the coefficients W_N. As can be seen from Fig. 7.2 and Appendix C.7, all the multiplications in the first stage and the first group of the other stages are by 1 (W_N^0) and therefore can be removed. Write an improved DIT FFT program to incorporate this enhancement. Count the number of cycles required to compute a 128-point transform and compare with the result in 1.

7A.4 The bit-reversed addressing mode provides a very efficient way to perform the bit-reversal and save execution time. In order to see its effect, write a program to explicitly perform the bit-reversal by swapping data elements in memory instead of using the bit-reversed addressing mode. Count the number of cycles required to perform this task and compare with the number of cycles required to compute the entire 128-point FFT.

7A.5 The program in Fig. 7.4 is a decimation-in-time algorithm. Change it to perform a decimation-in-frequency algorithm instead.

7.3 Real-time FFTs

As we mentioned earlier, the program in Fig. 7.4 only performs the FFT itself, it does not acquire data from the A/D converter and does not *do anything* with the resulting FFT coefficients. We will next discuss how to implement a real-time FFT and power spectrum analyzer using the TMS320C30 and the host PC. The implementation involves the following four aspects, which must be done continuously to produce a coherent frequency spectrum:

- Input a time waveform (real) and compute the FFT of the input samples.
- Calculate the power spectrum by squaring the real and imaginary parts of the FFT and sum them (Note: We have left out the square root for computational reasons).
- Copy the power spectrum data, as a function of frequency, to the PC via the TBC register.
- Display the frequency spectrum on the PC.

Since the input data to the FFT is real, our first task is to change the program to compute the DFT of a real sequence. There are several different ways to do that. The simplest approach, which we will use here, is to expand the real sequence with a zero imaginary part to obtain a complex sequence that can be used by the program in Fig. 7.4. This method, however, is also very inefficient because it requires approximately twice the number of operations and memory locations as the more optimal approaches discussed in Section 7.4.

The key point in this experiment is to carry out the data input, the data output, and the FFT calculation both simultaneously and continuously (well nearly anyway). To do this we break time up into frames of size N/F_s, where N is the FFT size and F_s is the sampling frequency. Within each frame, we need three intermediate data buffers. One buffer is used as an input buffer to receive samples from the A/D converter. The second buffer is the FFT buffer that contains N (say 128) samples on which the FFT is currently being computed. The third buffer is an output buffer for storing the previously calculated FFT points, which are ready to be sent to the D/A converter.

The program is best organized as a small interrupt-driven part that handles the input/output, and a main program (non interrupt-driven), which calculates the FFT. The reason for this division is that the I/O must be handled at the sample rate, while the FFT calculation is done at the frame-rate. At each sample instant, an interrupt occurs that prompts the I/O program to copy a sample from the A/D port into the input buffer and transmits a sample from the output buffer to the D/A port. After servicing the interrupt, control is returned to the main program, which will continue its FFT computation. Synchronization between the two parts is achieved by a single flag, which is set by the I/O program when the input buffer is full (output buffer is empty as well) and cleared by the main program once the FFT is under way. When the I/O program sets the flag, the FFT computation will also be finished and we can also switch the three buffers. We use the former output buffer to receive the input samples as the new input buffer. The former FFT buffer contains the new output data and becomes the output buffer. The original input buffer with N input samples becomes the current FFT buffer where the next FFT computation will take place. In this way, the three buffers work in a rotating pattern. Notice that we do not physically move any data around, but merely changes three pointers that control the purpose of the three buffers. Also notice that for this method to work properly, we assume that the FFT program runs fast enough to complete the FFT computation within the frame-time (N/F_s) (Why?). This should normally not be a problem with a sampling frequency of 8 kHz, unless N is very large.

Using this method, we can constantly read-in input samples, compute the FFT of a block of data, and output the frequency samples to the oscilloscope. On the oscilloscope, we get a constant display of the power spectrum of the input time signal. For the sweep on the oscilloscope to synchronize properly to the output sample points, we may need to supply a trigger pulse to the oscilloscope prior to sending each block of output data. Notice, however, that this trigger signal will distort the spectrum because of the D/A filter.

Figure 7.5 shows a main program for the spectrum analyzer. Notice that, as described above, the FFT calculation is done in the main program and not in the `receive0` interrupt service routine. As the main program is no longer just the dummy `idle` statement, it is *important* that the registers are saved (`push`ed) when entering the interrupt service routine and restored (`pop`ed) when leaving again, such that the main program is not disrupted by the interrupt service routine.

The `start_dma` routine, which can be found in `move.asm` in Appendix D.2, transfers the computed power spectrum values to the PC, where the `FFT_disp.c` routine (Appendix D.3) grabs them and displays them on the PC screen. The FFT spectrum values are also transferred to the D/A port, where you can observe the spectrum using an oscilloscope. Because of the filters in the D/A converter, however, the output resolution is rather poor, so this is for debugging purposes only.

```
*******************************************************************************
*                                                                             *
*                        FFT main Program - FFT_MAIN.ASM                      *
*                        -------------------------------                      *
*                                                                             *
*   Real-time FFT program that computes the DFT of a real-time input signal   *
*   and displays the power spectrum on oscilloscope and PC screen.            *
*                                                                             *
*      Henrik Sorensen.                                                       *
*      Written in Aug. 12, 1992. Modified in Feb 10, 1995. Version 1.0        *
```

```
*                                                                          *
*  Do not distribute without permission from authors. Copyright 1995-96    *
*                                                                          *
****************************************************************************
           .global  start
           .global  evm_init
           .global  receive0
           .global  p0_addr
           .global  stack_addr
           .global  start_dma

           .global  N
           .global  M
           .global  fft                    ;entry point of FFT codes
           .global  fft_buf

           .sect    "address"
buf_addr   .word    0809800h
           .word    0809900h
           .word    0809a00h

N          .set     128                    ;N=FFT length
M          .set     7                      ;M=log2(N)

           .text

bufaddr    .word    buf_addr
fft_size   .word    N
ipt_buf    .word    0809800h
fft_buf    .word    0809900h
opt_buf    .word    0809a00h
counter    .word    0

start: ldi     @stack_addr, sp      ;load stack address into stack pointer
       call    evm_init             ;initialize evm board
       ldi     @bufaddr, ar6
       ldi     3, bk

compu_fft:
       push    ar6                  ;save  ar6
       call    fft                  ;calculate FFT
       pop     ar6                  ;restore ar6

       ldi     @fft_buf, ar0        ;compute fft power spectrum
       ldi     ar0, ar1
       ldi     @fft_size, rc
       subi    1,rc
       rptb    power
       mpyf    *ar0, *ar0++, r0     ;X*X
       mpyf    *ar0, *ar0++, r1     ;Y*Y
       addf    r1,r0
       fix     r0,r0
```

```
        ash     -16,r0
power:  sti     r0,*ar1++

wait:   idle                        ;wait for interrupt
        ldi     @counter, ar7
        cmpi    @fft_size, ar7
        bne     wait                ;if buffer is not full, wait some more
        ldi     0, ar7
        sti     ar7, @counter       ;reset counter
        nop     *ar6--%             ;switch buffers
        ldi     *ar6++%, ar1        ;   ar1 -> new input buffer
        ldi     *ar6++%, ar0        ;   ar0 -> new FFT buffer
        ldi     *ar6++%, ar2        ;   ar2 -> new output buffer
        sti     ar1, @ipt_buf
        sti     ar0, @fft_buf
        sti     ar2, @opt_buf       ;store buffer pointers

* send data to host pc
        ldi     ar2, r6
        ldi     @fft_size, r7
        call    start_dma           ;subroutine that does transfer

        br      compu_fft           ;start new FFT calculation

receive0:                           ;interrupt service routine
        push    st                  ;save registers
        push    r0
        pushf   r0
        push    ar5
        push    ar7
        push    ar1
        push    ar2

        ldi     @p0_addr, ar5
        ldi     *+ar5(12), r0       ;read input sample
        lsh     16, r0              ;format input data
        ash     -18, r0
        float   r0, r0
        mpyf    0.1, r0             ;scale input sample
        ldi     @ipt_buf, ar1
        stf     r0, *ar1++          ;store input sample
        ldf     0.0, r0
        stf     r0, *ar1++          ;zero-expand imaginary part
        sti     ar1, @ipt_buf

        ldi     @opt_buf, ar2       ;send FFT spectrum value to D/A
        ldi     *ar2++, r0          ;load output sample
        sti     ar2, @opt_buf
        lsh     2, r0               ;get transmit word format
        sti     r0, *+ar5(8)        ;send output sample

        ldi     @counter, ar7       ;increment sample counter
```

```
nop     *++ar7
sti     ar7, @counter

pop     ar2                 ;interrupt service routine finished
pop     ar1                 ;restore registers
pop     ar7
pop     ar5
popf    r0
pop     r0
pop     st

reti                        ;return from interrupt
.end
```

Figure 7.5: FFT main program for TMS320C30

7.3.1 Experiment 7B: Real-time FFT and Power Spectrum Display

7B.1 Create a `fft_main.cmd` file and compile, link, and run the FFT spectrum analyzer. Use a sinusoidal sweep input and discuss how well the sinusoid is resolved. What are some of the limiting factors?

7B.2 Use a speech and an audio signal as input and observe the spectrum. How do the bandwidth of the two signals compare?

7B.3 Change the sampling frequency to 4 and 16 kHz, respectively. What is the bandwidth of the spectrum analyzer in these cases? Compare the FFT spectrum for different sampling rates.

7B.4 Use the real-time spectrum analyzer to find the spectrum of the random number generator that you implemented in experiment 4C. Observe and comment on its spectrum displayed on the oscilloscope. Compare the results with a commercial spectrum analyzer.

7B.5 Change the spectrum analyzer to use FFT of length 64 and 256, respectively. Compare the FFT output for the different lengths.

7B.6 The frequency spectrum is actually not being displayed continuously, but is displayed once per frame-time. This means that the displayed spectrum is actually sampled. What is that sampling frequency? Does aliasing occur?

7.4 Real-input FFTs

When the input to an FFT is real, the output will have inherent symmetries that can be used to reduce the number of operations to about half that required by a complex input FFT. Let $z[n] = x[n] + jy[n]$, where $x[n]$ and $y[n]$ are two real sequences. Compute the FFT of $z[n]$ to obtain $Z[k] = Z_r[k] + jZ_i[k]$, where $Z_r[k]$ and $Z_i[k]$ are the real and imaginary parts of $Z[k]$, respectively. The two individual transforms can be recovered as [13]

$$X[k] = \frac{1}{2}(Z_r[k]+Z_r[N-k])+j\frac{1}{2}(Z_i[k]-Z_i[N-k]) \quad k = 0, 1, \ldots \frac{N}{2} \tag{7.13}$$

$$Y[k] = \frac{1}{2}(Z_i[k]+Z_i[N-k])+j\frac{1}{2}(Z_r[N-k]-Z_r[k]) \quad k = 0, 1, \ldots \frac{N}{2} \tag{7.14}$$

An even more efficient real input FFT algorithm can be developed by exploiting the symmetries of the DFT represented for real data at every stage. A detailed discussion about the FFTs for real data can be found in [13].

7.4.1 Experiment 7C: Implementation of Real-input FFTs

7C.1 Modify your length-128 real-time complex-input FFT program to compute two real DFTs simultaneously. This will require extra buffering (and subsequent introduce extra delay) but will lower the computations per output point.

7C.2 Count the number of cycles to compute *one* length-128 *real* input FFT and compare it to the number of cycles for *one* length-128 *complex* input FFT.

7C.3 Modify the DIT FFT program in Fig. 7.4 to take advantage of the symmetries at every stage to develop the most efficient real-input radix-2 FFT (You might want to consult [13]). Count the number of cycles to calculate *one* length-128 *real* FFT and compare to the number you got in [2].

7.5 Hartley Transform

The Hartley transform is defined as

$$X[k] = \sum_{n=0}^{N-1} x[n]\text{cas}_N(nk) \qquad k = 0, 1, \ldots, N-1 \tag{7.15}$$

where $\text{cas}_N(x) = \cos(\frac{2\pi}{N}x) + \sin(\frac{2\pi}{N}x)$, which is the sum of the real and imaginary parts of the Fourier kernel.

7.5.1 Experiment 7D: Fast Hartley Transform Algorithms

7D.1 Write a Hartley transform algorithm by modifying your real-input FFT from 7C.3.

7D.2 Count the cycles required to compute a length-128 Hartley transform and compare it with the number of cycles required to compute a length-128 real input FFT.

7D.3 Use the oscilloscope to time your FFT and Hartley routines. Let your routine run in an infinite loop and at the start of your routine, transfer the value 128 to the output port. At the end of your routine, transfer -128 to the output port. Now you can hook up an oscilloscope and time the difference between start and end. Notice that due to the D/A filter, the output will not be square pulses as you might expect.

Chapter 8

Quantization Noise

8.1 Quantization

In this chapter, we will study the effects of quantization in digital signal processing systems using the TMS320C30. In theoretical analysis of DSP systems, it is generally assumed that the signal values and system coefficients are represented and computed in infinite-precision arithmetic. However, when the digital signal processing system is actually implemented, signals and coefficients must be represented and calculated in some number system (most often the binary number system) that is always of finite precision. Thus, we must approximate the infinite-precision values by finite-precision numbers, i.e., quantize the numbers. The quantization causes errors in the signal representation and processing, which may influence the performance of the system severely.

The total quantization noise in the output comes from three main sources. First, the A/D converter quantizes an otherwise infinite resolution analog waveform to a fixed-point binary representation at the sample instant. Second, system constants and variables with theoretically infinite or very high (generally at least 32-bit floating-point) accuracy, but which must often be quantized to a more limited word-length. Finally, quantization is often needed or occurs implicitly during the arithmetic calculations. Even if the signals and coefficients can be represented without any quantization error, the result of a computation often exceeds the available word-length. For example, the multiplication of two 32-bit numbers will produce a 64-bit result, which must be quantized to the "closest" 32-bit number before it can be stored back to memory.

Since the quantization noise comes from calculating and representing the variables in a finite word-length representation, it is obvious that the shorter the word-length used, the more severe the quantization errors will be. For a given word-length, the total quantization noise also depends on the specific function being computed. For example, quantization effects are generally more pronounced in an IIR filter than in a FIR filter because of the feedback in the IIR filter (Why?). The quantization noise is also affected by the different structural realizations. The direct form structure for IIR filters, for example, is more sensitive to quantization errors than the cascade structure [8, 9].

The possibility of overflow is another problem to overcome when DSP functions are implemented on a fixed-point device, like the TMS320C1X or the TMS320C2X, due to their limited dynamic range. The input and intermediate signals must be scaled properly to avoid overflow while still maintaining a reasonable dynamic range at all times. Notice, whereas normal quantization noise causes fairly small errors (around 2^{-B} with a B bit word length), an overflow creates much larger errors; especially if saturation logic is not used. The scaling factors should, hence, be chosen as a trade-off between a large dynamic range (i.e. small quantization noise) and a low overflow-noise (i.e. larger quantization noise). Implementing a system in floating-point arithmetic greatly decreases the problems of overflow and quantization effects. By representing the exponential and fractional component of each sample separately, the floating-point arithmetic can automatically achieve the optimal scaling within a large dynamic range and, therefore, achieve an overall low quantization noise.

The zero-input limit cycle behavior is another phenomenon that can occur when implementing a feedback system (such as an IIR filter) with finite-precision arithmetic. The output of the system may continue to oscillate indefinitely while the input remains zero. This can happen due to the quantization errors (small scale limit cycles) or due to overflow errors (large scale limit cycles).

A detailed discussion of the finite-precision numerical effects can be found in [8, 9].

8.2 Study of Quantization Effects

Because of the large dynamic range of the TMS320C30 (32/40-bits), it can be difficult to detect the effects of quantization. Since our purpose is to study these effects, the implementations discussed here are not meant as a "realistic" use of the TMS320C30, but will exaggerate the the quantization errors to demonstrate its effects using different word-lengths, filter structures, and both types of arithmetic.

As we know, the A/D converter on the TMS320C30 EVM board has 14 bits of dynamic range, which includes one sign bit. This corresponds to a signal-to-noise ratio (SNR) around 83 dB such that the original signal is well represented by the sampled data. To make the quantization effects more discernable, we need to reduce the bits of the A/D conversion. This can be simulated on the TMS320C30 by masking (`AND`ing) bits off the conversion result before using it in any calculations. We will write a program that reads a sample from the A/D converter, masks (`AND`s) certain bits off by setting them to zero, and then outputs the masked sample to the D/A converter. If the mask is made a global variable, it can easily be changed in the debugger to simulate any number of bits in the A/D conversion below 14.

With the 32-bit floating-point arithmetic of the TMS320C30, the problems of quantization effects and overflow are almost eliminated, as can be seen from our previous experiments. However, quantization effects still exist in floating-point arithmetic and they increase with a decreasing mantissa length. To more readily observe the quantization noise using floating-point arithmetic, we will change the length of the mantissa. The TMS320C30's floating-point arithmetic uses 24 bits for the mantissa. We can reduce the length by using the masking (`AND`ing) technique mentioned above. Notice that to simulate a real 16-bit or 8-bit mantissa length, both the coefficients and the registers must be bits-masked properly to match the desired mantissa length.

To study the quantization effects of fixed-point arithmetic, we will also implement some DSP functions in fixed-point arithmetic. Fixed-point instructions are available on the TMS320C30, but are generally only used for address computations. Unlike the case of floating-point arithmetic, where we can use floating-point coefficients directly and implement the functions directly from their formula without any other considerations, the implementation in fixed-point arithmetic is more involved. First, the coefficients must be represented as integers. This can be done by multiplying all coefficients, which are generally small decimals, by some large number M and round them to the closest integer. The multiplicative factor M should be chosen properly so that the resulting integer coefficients are represented with adequate precision while the largest coefficient is still less than the maximum integer we have available for the given word length. Second, we have to make sure that the intermediate results at any node do not exceed the maximum integer and, hence, cause overflow.

8.2.1 Experiment 8A: Quantization Noise in Sampling

8A.1 Modify the sampling program in Experiment 3A to mask off bits of the input samples and then output the masked samples. Alter the mask to obtain an output corresponding to 14, 12, 10, 8, 6, and 4-bit quantization. Set the amplitude of the input sinewave signal to about 1.5 V (such that the samples from the A/D converter have the full 14-bit resolution). Observe the output on the oscilloscope. Record the point at which the quantization effects become visible. What SNR does that correspond to?

8A.2 Connect a microphone and speaker to the EVM. Repeat the steps in 3A.1. Listen to the quantization effects to the speech signal. Record the point at which the quantization effects become detectable. What SNR does that correspond to?

8A.3 Calculate the theoretical mean and variance of the noise for each of the cases above, estimate the values obtained from your oscilloscope, and compare the two values.

8.2.2 Experiment 8B: Quantization Noise with Floating-point Arithmetic

8B.1 Modify the FIR lowpass filter program in Fig. 5.3 to mask off bits of the mantissa. Change the mantissa length from 24, to 16, to 8 bits. Make sure that both the coefficients and registers are bits-masked properly. Observe the filter response from the oscilloscope and record the results.

8B.2 Modify the IIR lowpass filter program in Fig. 6.4 to mask off bits of mantissa. Follow the same steps as above to observe the quantization effects in the IIR filter. Compare the results with the FIR filter.

8.2.3 Experiment 8C: Quantization Noise in Fixed-point Arithmetic

8C.1 Modify the FIR lowpass filter in Fig. 5.3 to use fixed-point arithmetic. The filter coefficients must be represented as integers and the problem of overflow should be considered. The fixed-point arithmetic of the TMS320C30 is 32 bits. Use the mask-off method to achieve a 32-bit, 24-bit, and 16-bit arithmetic. Make sure to mask off bits correctly so that you really are using 24-bit or 16-bit word length and arithmetic. Observe the filter responses for the different word lengths and record the results.

8C.2 Modify the program in Fig. 6.4 to implement the order-4 lowpass

IIR filter in fixed-point arithmetic. You can use either the direct form or the cascade structure. Before the implementation, we suggest that you study the filter response (system function) and determine a scaling strategy. As you did above, use the mask-off method to realize 32-bit, 24-bit, and 16-bit word-lengths arithmetic, respectively. Observe the quantization effects in the IIR filter. Compare the results with the FIR filter.

8.2.4 Experiment 8D: Limit Cycles

8D.1 Observe large scale limit cycles by increasing the input amplitude in the fixed-point IIR filter above, until overflow occurs. Some very large tones will occur, which are called limit cycles.

8D.2 Change the fixed-point IIR filter to implore saturation logic, and repeat the experiment above.

8D.3 Observe small scale limit cycles by turning off the signal source and watch the periodic small signal residue on the output. Notice that it will not always occur (depends on the state of the filter when the input disappears), so try a couple of times to see them.

Chapter 9

Adaptive Filters

9.1 Adaptive Filters

In Chapter 5 and 6, we described the implementation of FIR and IIR filters, where the filter coefficients were designed according to a given set of specifications. In many applications, however, the filter coefficients can not be specified a priori because of uncertainty about the signal or because of an undetermined design criteria due to a changing environment. An example is a speaker-phone echo canceler, where the desired output mimics the echo part of the input signal, such that it can be subtracted from the input signal to produce an "echo-free" signal. The filter coefficients can not be determined ahead of time since they depend on the room acoustics. For applications such as this, it is necessary to rely on adaptive filtering techniques.

An adaptive filter is an adjustable filter where the coefficients are updated at regular intervals. The updates are determined by an adaptive process that tries to optimize the filter response with respect to some performance criterion. The filter, which performs the signal modifications, and the adaptive algorithm are generally two distinct parts of the adaptive filter, as illustrated in Fig. 9.1.

Although both FIR and IIR filters can be considered for the filter part, the FIR filter is by far the most practical and widely used. This is because IIR filters have poles that can cause instabilities as the filter coefficients are updated, while FIR filters only have zeros—and, hence, are stable irrespective of the filter coefficients. Of the various FIR filter structures that we may use, the Direct Form and the Lattice Form are the two most frequently used in adaptive filtering applications.

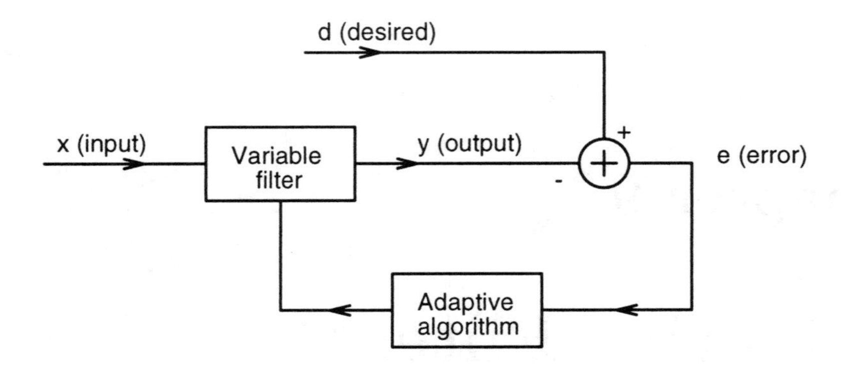

Figure 9.1: General Form of an Adaptive Filter

Another important consideration in the use of an adaptive filter is the choice of an adaptation algorithm for adjusting the filter coefficients. Much of the reported research on adaptive filters have been based on Widrow's well-known Least Mean Square (LMS) algorithm [21]. The LMS algorithm is relatively simple to design and implement, and is well-suited for many applications. This algorithm, combined with the Direct Form FIR structure, will be used in our experiments.

The Direct Form FIR structure was discussed in Chapter 5. A detailed description of the LMS algorithm can be found in many books on adaptive filters such as [21]. In this algorithm, an error signal

$$e[n] = d[n] - y[n] \tag{9.1}$$

is used to adjust the filter coefficients, where $d[n]$ is the desired filter output and $y[n]$ is the actual filter output. The filter coefficients are updated according to the equation (see Eqn. (6.3) in [21])

$$h[n+1,k] = h[n,k] + 2 * \mu * e[n] * x[n-k] \qquad k = 0, 1, \ldots, N-1 \tag{9.2}$$

where μ is a constant called the step size, $x[n]$ is the input signal, and $h[n,k]$ is the kth filter coefficient at time n.

The step size μ controls the convergence rate of the algorithm. A large value of μ leads to large step size adjustments and, thus, to rapid convergence. However, if μ is made too large, the algorithm does not converge. To ensure convergence, μ must be chosen in the range (see Eqn. (6.10) in [21])

$$0 < \mu < \frac{1}{N * P_x} \tag{9.3}$$

where N is the length of the adaptive FIR filter and P_x is the power in the

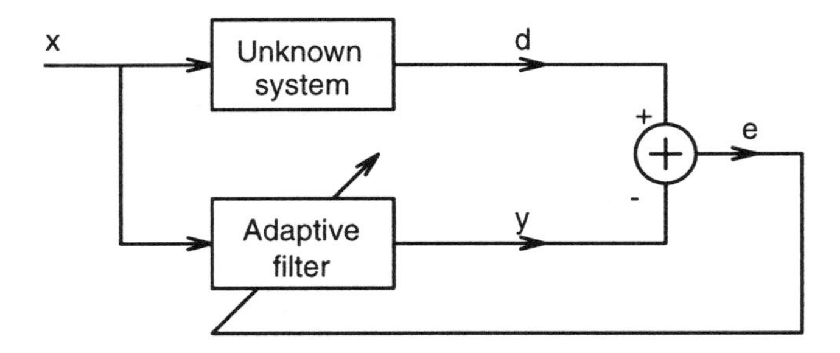

Figure 9.2: System Identification

input signal. P_x can be approximated by

$$P_x = \frac{1}{M+1}\sum_{n=0}^{M} x[n]^2 \tag{9.4}$$

where M is a constant.

This algorithm is actually a recursive gradient (steepest-descent) method. When μ is selected within the range given by Eqn. (9.3) and the initial values, $h[0, k] = 0$ for $0 \leq k < N$, then the coefficients will converge to their optimal solution, if they are updated by Eqn. (9.2) and provided the input is stationary.

Adaptive filters are used in various applications such as adaptive prediction, channel equalization, system identification, echo cancelation, and noise cancelation. In this chapter, we will use system identification and noise cancelation applications to develop our experiments.

The adaptive approach is very useful in system identification or modeling, and the structure used is shown in Fig. 9.2. The output $d[n]$ of an unknown system with input $x[n]$ is the desired signal. The response of the adaptive filter $y[n]$, to the same input $x[n]$, is compared to the unknown system output $d[n]$ to create the error signal $e[n]$, which is used to adjust the adaptive filter. The filter will be adjusted until the error signal reaches approximately zero. At this time, the adaptation is complete and the behavior of the resulting filter matches the unknown system; hence, it is a model of the unknown system. In this scheme, the input $x[n]$ should be bandlimited white noise with a spectrum broad enough to excite all the poles and zeros of the unknown system.

Shown in Fig. 9.3 is the diagram of a noise canceler. Here, the the measured signal, $d[n]$, is the desired signal, $x[n]$, contaminated with noise, $s[n]$. The filter input in this case is $s'[n]$, which is the noise measurement that will be similar to $s[n]$ but will not necessarily equal $s[n]$ exactly. $s'[n]$ should NOT contain any of the $x[n]$ signal component. This usually means

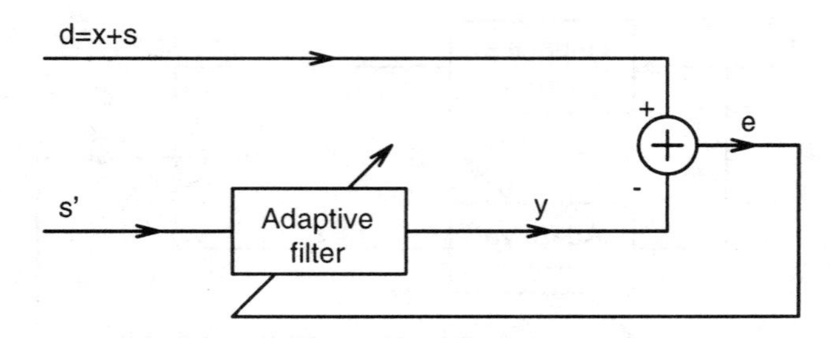

Figure 9.3: Noise Cancelation

that $s[n]$ and $s'[n]$ come from the same source, but that $s'[n]$ has been modified by the environment in some way. An adaptive filter is used to estimate the noise in $d[n]$ and the noise estimate $y[n]$ is subtracted from $d[n]$ to produce $e[n]$. If $x[n]$ is uncorrelated with $s[n]$, the filter output $y[n]$ will approach the noise $s[n]$. The difference $d[n] - y[n]$ will thus approach $x[n]$ and we have a noise canceler.

9.2 Implementation of Adaptive Filters

The adaptive filter shown in Fig. 9.2 consists of four basic modules as shown in Fig. 9.4: an unknown system with the desired output $d[n]$, an adaptive FIR filter with the adjusted response $y[n]$, a module that updates the coefficients using the LMS algorithm, and a white noise signal for the input $x[n]$. In normal use, all these modules, with the exception of the unknown system, will be implemented on the DSP as illustrated in Fig. 9.4. To allow better insight into the adaptation process and to avoid building hardware, we will also, for some experiments, implement the "unknown system" on the DSP. Since we can adapt to both poles and zeros, we can use any of the FIR and IIR filters from Chapter 5 and 6, respectively, as the unknown system.

The Direct Form FIR structure is available from Chapter 5, except that the coefficients are changed every time a new sample arrives. Starting with zeros, the coefficients are updated according to Eqn. (9.2), which is easy to implement. An initial step size μ is chosen to satisfy Eqn. (9.3) and the actual value can be determined by trial and error during the debugging phase.

The noise source, $x[n]$, can be generated internally on the TMS320C30 based on the random number generator that was discussed in Section 4.3 and the code developed in the corresponding experiments. However, when the "unknown system" is implemented internally on the DSP, the white

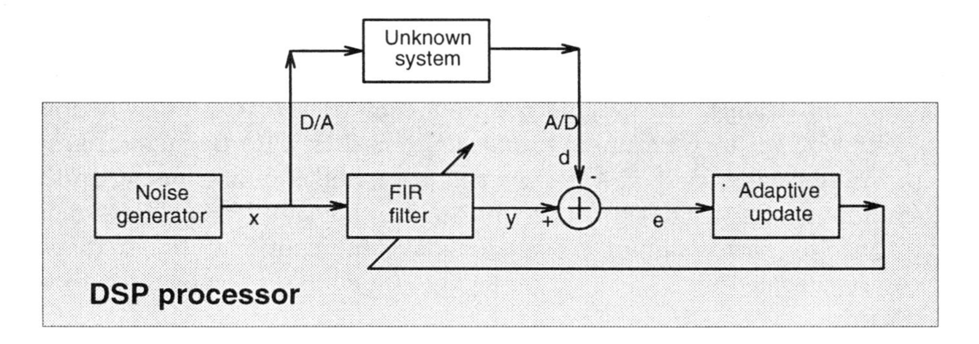

Figure 9.4: Normal setup for system identification applications

noise signal may instead come from an external noise generator. Some experiments will even employ a sweeping sinusoidal signal to emulate the "noise source." When the input to the system and the adaptive filter consists of a single frequency, the filter will adapt to match the unknown system at that particular frequency. If we now sweep the input frequency wide and fast enough, the resulting adaptive filter will approach the unknown system at all the frequencies and the coefficients will convergence to the optimal solution. The advantage of using a sinusoidal signal is that we can easily observe the adaptation process since the frequency selective characteristics of the system with a sweeping sinewave input can be viewed directly on an oscilloscope. By connecting the output of the unknown system and the adaptive filter to the oscilloscope, we can observe their frequency responses and compare them with each other (Notice on the EVM board you can only view one output at a time since there is only one D/A port). On the other hand, it is also an interesting experiment to use a single frequency sinewave input and examine the resulting filter coefficients and compare them with the unknown system. However, in all practical applications, we should certainly use a real white noise signal.

To perform noise cancelation on the system shown in Fig. 9.3, we need access to the three signal $x[n]$, $s[n]$, and $s'[n]$. $x[n]$ is the signal that we want to extract from the noisy version. $s[n]$ is a noise signal. $x[n]$ plus $s[n]$ forms the noise-contained signal that serves as the comparison signal for the adaptive structure. $s'[n]$ is the input to the adaptive filter. It is correlated to $s[n]$ but somehow different from $s[n]$ in a practical application. In this experiment, we want to focus on the adaptive part and will hence use $s'[n]$ equal to $s[n]$ since it also makes the system simpler. We will use sinewaves as the signal $x[n]$ since it is easy to identify on an oscilloscope as discussed earlier. The noise signal $s[n]$ can be chosen as bandlimited noise, sinewave, or squarewave.

Of the two signals, $x[n]$ and $s[n]$, one must be generated internally on the TMS320C30; the other may come from an external function generator. For the convenience of changing the frequency or the shape of the noise, it is often better to generate $s[n]$ externally and let $x[n]$ be generated on-chip. We normally want $x[n]$ to be a sinewave, and the internal generation of sinewaves was discussed in Section 4.2.2. The implementation of the adaptive filter part is similar to the one for system identification. To display the noise-canceled output, the error signal $e[n]$ should be connected to the oscilloscope.

9.2.1 Experiment 9A: Adaptive Filter for System Identification

9A.1 According to the above discussion, write a program to implement the adaptive system identification shown in Fig. 9.2. Use the lowpass FIR filter of order-25 in Table 5.1 as the unknown system. Choose the order of the adaptive filter to be the same as that of the unknown system. Use a sinewave signal as the input and connect the output of the adaptive filter, $y[n]$, to an oscilloscope.
Adjust the input frequency and observe the output on the oscilloscope. If your program works correctly, you will see a lowpass response that matches the unknown system. You can connect the output of the unknown system, $d[n]$, to the oscilloscope to compare.
Let the input frequency sweep from 0 to 4 kHz and run the program. After several sweeps, halt the execution. On the debugger, check the converged coefficients of the adaptive filter and see how they approach the coefficients of the unknown system.
Run your program again with a single frequency as the noise source and verify that the resulting coefficients differ from the ones of the unknown system.
Reload your program and use a "sweeping sinewave" as noise. During the first several sweeps, halt the execution a few times and examine how the adaptive coefficients approach the coefficients of the unknown system step by step. (You can actually modify your program slightly and output the coefficient error to the oscilloscope).

9A.2 Increase and reduce, respectively, the order of the adaptive filter and observe the output. With a sweeping input, examine the adapted coefficients.

9A.3 Modify your program to use the lowpass IIR filter of order 4 in Experiment 6A.1 as the unknown system. Observe the adaptive output and examine the converged coefficients.

9A.4 Build (or borrow) a hardware second order Butterworth filter with a cut-off of 2 kHz. Use this filter as the unknown system as in Fig. 9.4. Use white noise as input and watch the filter coefficients adapt. Is the resulting filter a Butterworth filter with cutt-off of 2 kHz?

9.2.2 Experiment 9B: Adaptive Filter for Noise Cancelation

9B.1 Write a program to implement the noise cancelation system shown in Fig. 9.3. Generate a sinewave internally as the actual signal, $x[n]$, and use another sinusoidal signal from a function generator as the noise. As discussed in Section 9.2, you can use the noise $s[n]$ as the input for the adaptive filter.
First, output the "measured signal" (i.e., $x[n] + s[n]$) to the D/A converter and observe it using an oscilloscope to see how the contaminated signal appears. Next, output the error signal, $e[n]$, to the D/A converter and observe via the oscilloscope. If your program works correctly, you will obtain a noise-canceled sinewave signal that matches the useful input.
Adjust or sweep the frequency of the noise signal to see that the adaptation is achieved at any frequency and the desired sinewave output is maintained.

9B.2 Alter the value of the step size μ to change the convergence speed and observe the adaptation process. With a small step size, you can clearly see the gradual noise elimination on an oscilloscope.

9B.3 Change the noise signal to a squarewave or a bandlimited noise signal to see how it affects the adaptation process.

Chapter 10

Multirate Signal Processing

10.1 Multirate Systems

Multirate signal processing systems use more than one sampling frequency to perform its DSP tasks. Applications for multirate signal processing can be found in many areas, including communications, speech, and audio processing. The fundamental operation of any multirate system is the sampling rate conversion—raising or lowering the sampling rate of a signal.

This chapter presents experiments for sampling rate reduction and increase as well as some experiments centered around a filterbank. Before describing the implementation of the experiments, an introduction is given to both sampling rate conversion and filter banks. Since multirate processing is not a central part of most digital signal processing courses, the introduction in this chapter is more detailed than for the other chapters in this book. For a more general discussion of multirate signal processing, refer to various books [3, 20].

10.1.1 Sampling Rate Reduction—Decimation by an Integer Factor

Assume that $x[n]$ represents a discrete-time sequence obtained by sampling the function $x_c(t)$ with sampling frequency F. Also assume that $x[n]$ is a full band signal in the range $\frac{-F}{2} \leq f \leq \frac{F}{2}$, or $-\pi \leq \omega \leq \pi$ with $\omega = 2\pi fT$, where T is the sampling period. Consider the process of reducing the sampling rate of $x[n]$ by an integer factor M. This can be achieved by "resampling" $x[n]$, i.e., by defining a new sequence

$$x_d[n] = x_c(nT') = x_c(nMT) = x[nM] \tag{10.1}$$

where the new sampling period, T', is

$$T' = MT \tag{10.2}$$

Then the new sampling rate is

$$F' = \frac{1}{T'} = \frac{1}{MT} = \frac{F}{M} \tag{10.3}$$

From the Nyquist sampling theorem, we know that aliasing can occur in the resampled signal because of the reduced sampling rate. In order to avoid aliasing, it is necessary to prefilter the signal $x[n]$ with a digital lowpass filter that has a cutoff frequency $\frac{F}{2M}$ or $\frac{\pi}{M}$. The sampling rate reduction is then achieved by saving only every Mth sample of the filtered output. This process is illustrated in Fig. 10.1(a). The filtered output is given by

$$w[n] = \sum_{k=-\infty}^{\infty} h[k]x[n-k] \tag{10.4}$$

where $h[n]$ denotes the impulse response of the lowpass filter. The final output $x_d[n]$ is then obtained as

$$x_d[n] = w[nM] \tag{10.5}$$

By combining Eqn. (10.4) and (10.5), the relation between the sample-rate reduced sequence, $x_d[n]$, and the original signal, $x[n]$, is given by

$$x_d[n] = \sum_{k=-\infty}^{\infty} h[k]x[nM-k] \tag{10.6}$$

In general, the second block in Fig. 10.1(a), which describes Eqn. (10.5), is referred to as a sampling rate compressor. The operation of reducing the sampling rate, including the prefiltering, is called down sampling or decimation. Fig. 10.1(b) shows the typical spectra of the signals $x[n]$, $h[n]$, $w[n]$, and $x_d[n]$ for the decimation by an integer-factor M.

From the proceeding discussion and the block diagram in Fig. 10.1(a), we see that it is not necessary to compute each output of the lowpass filter at the sampling rate F since only every Mth output is needed. When using a FIR filter, which is related only to the past inputs, the filter computation can be performed at the lower rate $F' = \frac{F}{M}$.

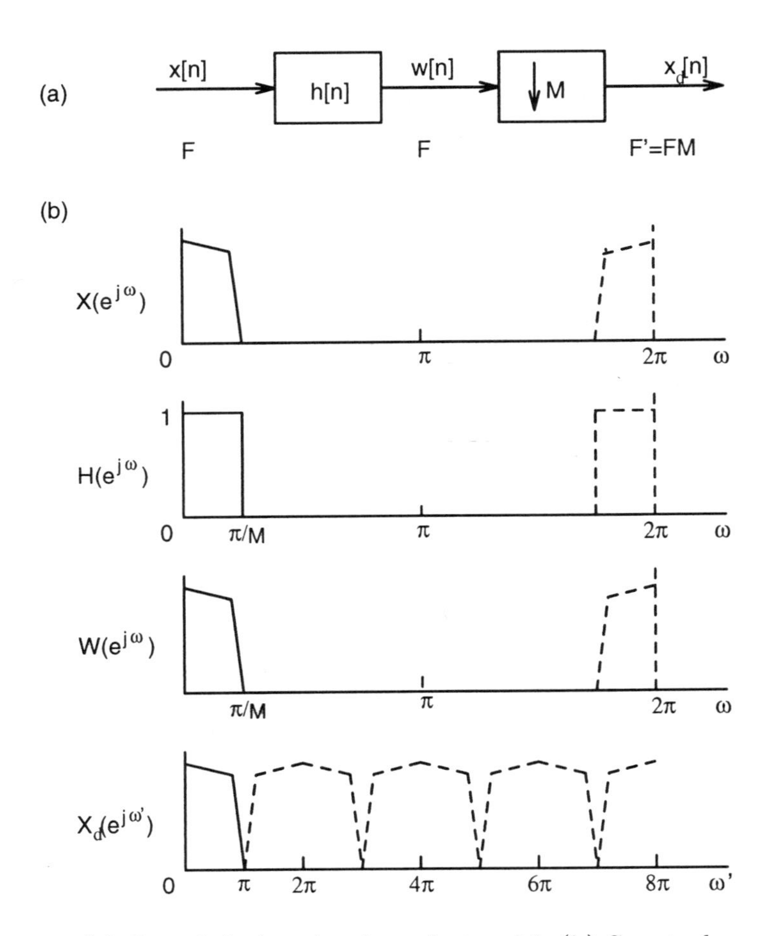

Figure 10.1: (a) Signal decimation by a factor M; (b) Spectral representation of decimation process

10.1.2 Sampling Rate Increase—Interpolation by an Integer Factor

Similar to the decimation, a sampling rate increase by an integer factor L can be obtained by defining a new sequence

$$x_i[n] = x_c(nT') = x_c(\frac{T}{L}) \tag{10.7}$$

where $T' = T/L$ is the new sampling period and the new sampling rate is

$$F' = LF \tag{10.8}$$

This process is generally called upsampling or interpolation.

An increase in the sampling rate of a signal $x[n]$ implies that we must interpolate $L-1$ new sample values between every two samples in $x[n]$. Fig. 10.2 illustrates an example of interpolation by a factor $L = 3$. The input signal $x[n]$ is first expanded with $L-1$ zero samples between every two samples of $x[n]$, resulting in

$$w[n] = \begin{cases} x[n/L] & n = 0, \pm L, \pm 2L, \ldots \\ 0 & otherwise \end{cases} \tag{10.9}$$

The first block in Fig. 10.2(a) denotes this operation and it is referred to as a sample rate expander. According to the sampling theorem, the spectrum of the input $x[n]$ contains not only the baseband frequencies of interest (i.e., $-F/2$ to $F/2$) but also images of the baseband centered at the harmonics of the sampling frequency, $\pm F, \pm 2F, \ldots$. When the sampling rate is increased to LF, the spectrum of the signal $w[n]$ has L images in the baseband region from 0 to $F'/2$, as shown in Fig. 10.2(c). To recover the signal $x[n]$, we must eliminate the unwanted image components by filtering the signal $w[n]$ with a digital lowpass filter that has a cutoff frequency $F/2$ or π/L (where $\omega = 2\pi f T'$). Fig. 10.2(a) shows the system for the sampling rate increase by a factor of L.

The output $x_i[n]$ from Fig. 10.2(a) can be expressed as

$$x_i[n] = \sum_{k=-\infty}^{\infty} h[n-k]w[k] \tag{10.10}$$

Combining Eqn. (10.9) and (10.10), we have the input-to-output relation of the interpolator

$$x_i[n] = \sum_{k=-\infty}^{\infty} h[n-k]x[k/L], \quad \frac{k}{L} \text{ an integer} \tag{10.11}$$

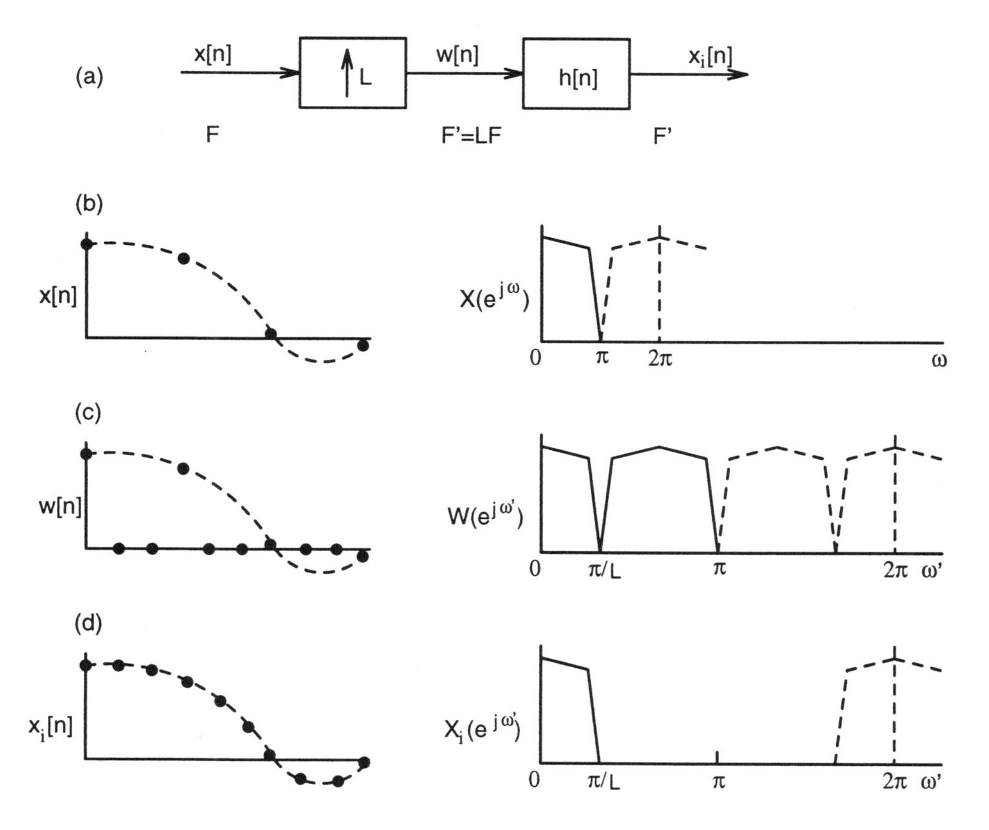

Figure 10.2: Block diagram and typical waveform and spectra for interpolation by an integer factor $L = 3$

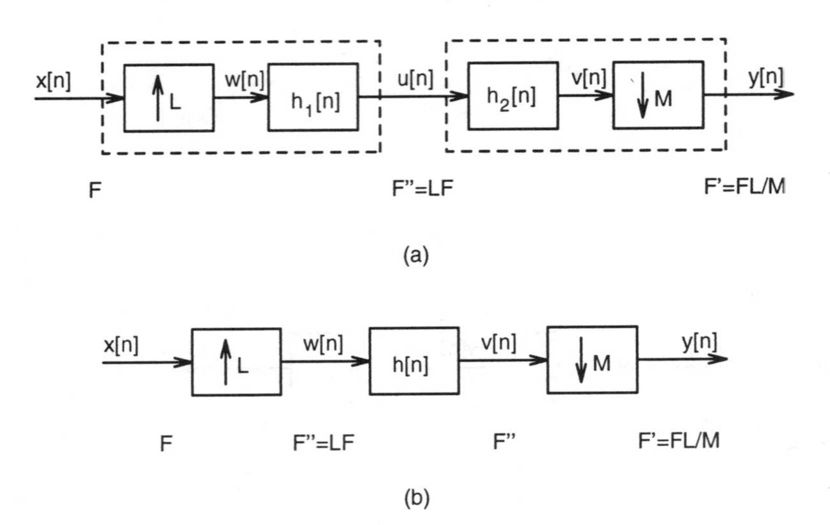

Figure 10.3: (a) System for changing sampling rate by a non-integer factor. (b) Simplified system in which the decimation and interpolation filters are combined.

$$= \sum_{k=-\infty}^{\infty} h[n - rL]x[r] \tag{10.12}$$

Finally, it should be noted that the lowpass filter must have a gain of L in the passband in order to ensure a correct amplitude for $x_i[n]$. This is due to the amplitude scaling in the resampling process.

Now that we have shown how to increase and decrease the sampling rate of a sequence by an integer factor, consider a more general case of changing the sampling rate by a noninteger factor by cascading an interpolation and a decimation module. Fig. 10.3 illustrates this process where it should be noted that the two low-pass filter can be merged, as shown in Fig. 10.3(b).

10.1.3 A Multirate System of Filter Banks

Digital filter banks are used, for example, in systems for speech analysis, bandwidth compression, and spectral parameterization of signals [3, 8]. These systems generally have some form of filter bank decomposition or reconstruction of a signal and involve the operations of sampling rate decimation and interpolation. Fig. 10.4 illustrates the basic framework for a K-channel filter bank analyzer and synthesizer. In the analyzer, the input signal $x[n]$ is divided into a set of K channels, $x_k[n]$ for $k = 0, 1, \ldots K - 1$, which occur in a decimated form. In the synthesizer, the K signals, which may have been modified depending on the application, are interpolated and

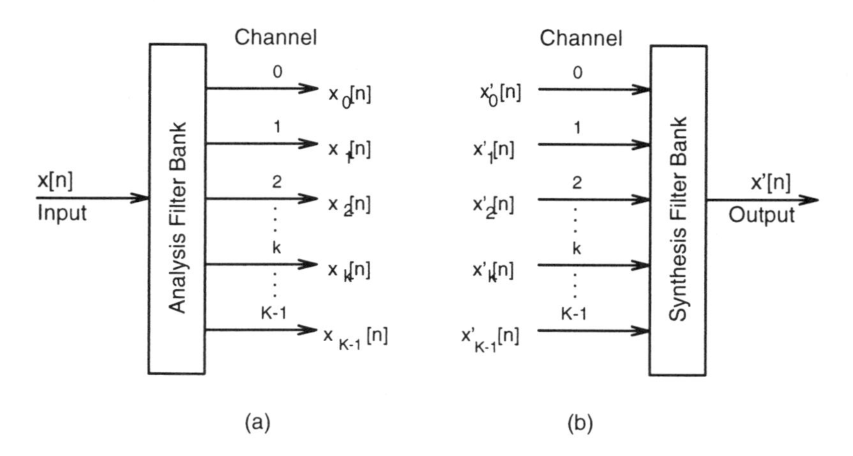

Figure 10.4: Basic framework for (a) a K-channel filter bank analyzer; and (b) a K-channel filter bank synthesizer

combined to reconstruct the signal $x'[n]$.

Each of the K channels is essentially a bandpass signal. To represent a bandpass signal, the sampling rate must be twice that of the highest-frequency component in the band. By translating (modulating) the band-pass signal to the lowpass band, however, the sampling rate necessary to represent this signal can be reduced to twice the channel bandwidth. Decimation and interpolation of a bandpass signal can be directly realized when dealing with integer-band channels. It is achieved by taking advantage of the inherent frequency aliasing or imaging properties of decimation and interpolation.

Fig. 10.5 illustrates this process of generating $x_2[n]$, the second out of M bands, from $x[n]$. The input signal $x[n]$ is first filtered by a bandpass filter $h_{BP}[n]$ to isolate the frequency band of interest. The sample-rate of the resulting bandpass signal, $x_{BP}[n]$, is then directly reduced by an M-sample compressor to produce the final output $x_2[n]$. As can be seen from Fig. 10.5(d), using the image of $x_{BP}(e^{j\omega})$ in the lowpass band (from 0 to F/M) $x_2[n]$ is a modulated version of $x_{BP}[n]$ when the sampling rate is reduced by a factor of M. By the inverse process of interpolation, the bandpass signal $x_{BP}[n]$ can be uniquely reconstructed from $x_k[n]$ ($k = 2$ in Fig. 10.5).

It is seen that the system in Fig. 10.5(a) is identical to that of the integer lowpass decimator discussed in Section 10.1.1, except that the filter is a bandpass filter instead of a lowpass filter. The processes of modulation and sampling rate reduction are achieved simultaneously by the M-to-1 compressor. Since the modulation is restricted to values of $2\pi i/M$, it is clear

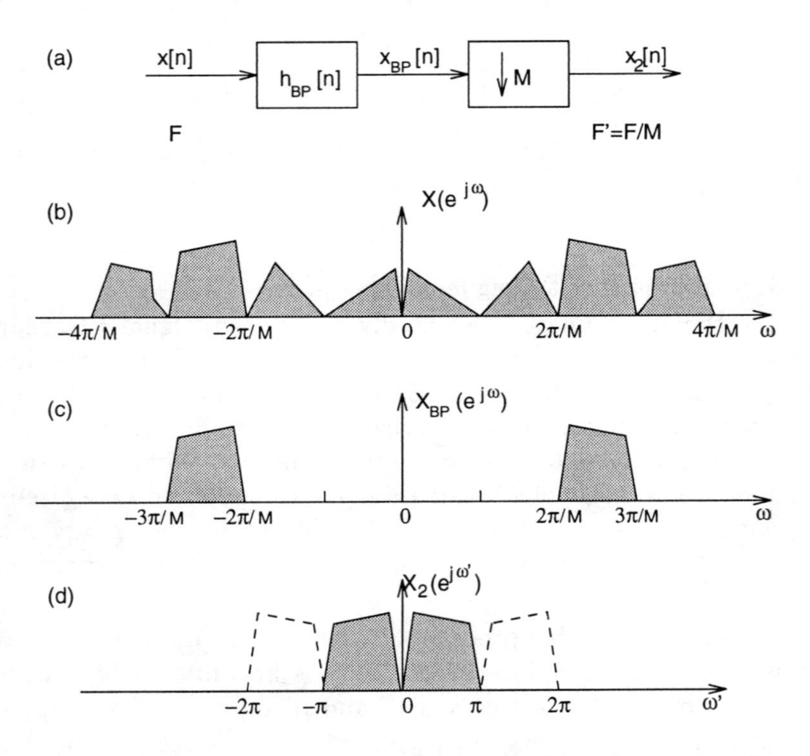

Figure 10.5: Integer-band decimation of bandpass signals

that only integer bands are allowed by this method; that is, the bands must be from $k\pi/M$ to $(k+1)\pi/M$ where k is an integer such that $0 \leq k \leq M-1$. If the constraints are not satisfied, nonrecoverable aliasing occurs in the baseband of $X_2(e^{j\omega'})$ and the signal $x_{BP}[n]$ cannot be reconstructed from $x_k[n]$.

It should also be noted that for all the odd-numbered bands (k odd), the spectrum is inverted in the process of lowpass translation, which can be seen from the example in Fig. 10.5, if the bandpass region of $h_{BP}[n]$ is $3\pi/M \leq \omega \leq 4\pi/M$. This is a consequence of the fact that even numbered bands correspond to upper sidebands of the modulation frequencies (kF/M) and odd numbered bands correspond to the lower sidebands, whereas the baseband of $X_k(e^{j\omega'})$ is of a upper sideband. The spectrum reversal might not cause any real problem, since the inverse interpolation process also inverts the spectrum of an odd-numbered band and the correct spectrum is, hence, recovered automatically.

The integer-band interpolation is the inverse process to that of integer-band decimation. It reconstructs (interpolates) a bandpass signal from its integer-band decimated representation. Fig. 10.6 illustrates this process. The sampling rate of the input signal $x[n]$ is expanded by a factor L by inserting $L-1$ zeros between each pair of samples to produce the signal $w[n]$. From the discussion of integer interpolation in Section 10.1.2, we know that the spectrum of $w[n]$ consists of periodically repeated images of the baseband of $X(e^{jw})$ centered at the harmonics $kF' = kF/L$. A bandpass filter is then used to select the appropriate image of this signal. Fig. 10.6(d) and (e) shows the output spectrum of the bandpass signal $X_k(e^{j\omega'})$ for the $k=2$ and $k=3$ band, respectively. As in the case of integer-band decimation, it is seen that the spectrum of the resulting bandpass signal is inverted for odd values of k.

Combining the ideas above, Fig. 10.7 illustrates a 2-channel filter bank system using the integer-band decimation and interpolation. After the decimation stage, the signals are quantized (Q) and multiplexed onto the channel for transmission, storage, etc. Using this system, we can conduct some experiments on speech coding. Subband coding is a technique used to efficiently encode speech signals below 64 kbits/sec. by taking advantage of the uneven distribution of energy in the speech spectrum. Breaking a speech signal into a number of frequency bands, it is possible to reduce the number of quantization levels for those bands with less energy. Assume that the input in Fig. 10.7 is a speech signal with a 4 kHz bandwidth (sampled at 8 kHz) and each sample is represented with 14 binary bits, which corresponds to a bit-rate of $14 \times 8 = 112$ kbits/sec. In the analyzer, the speech signal is split by filtering into two frequency channels, a lowpass band from 0 to 2 kHz and a highpass band from 2 kHz to 4 kHz. Each of the

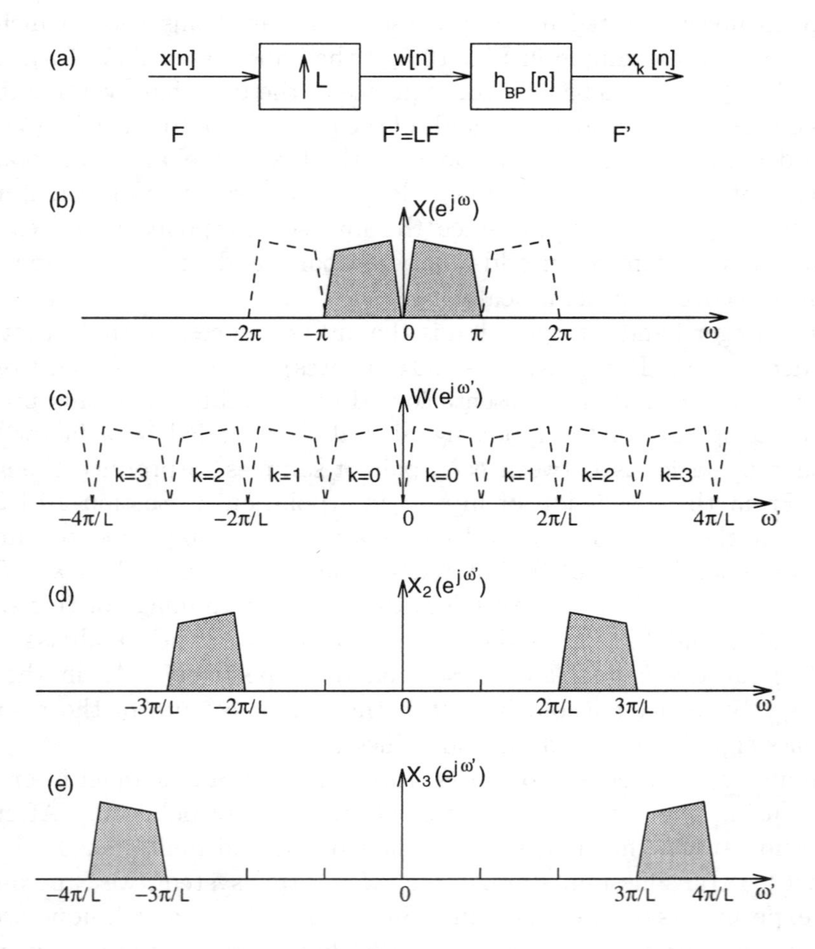

Figure 10.6: Integer-band interpolation of bandpass signals

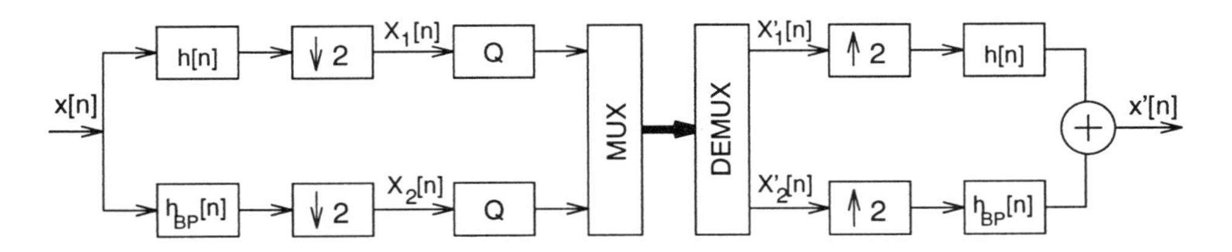

Figure 10.7: A 2-channel filter bank system used as a subband compressor

two bands is then sample rate reduced from 8 kHz to 4 kHz. If we quantize all samples to 7 bits, we end up with a channel bit-rate of $2 \times 7 \times 4 = 56$ kbits/sec. An alternate approach would be to use 8 bits to represent band one and 6 bits to represent band two, which results in the same channel bit-rate since $8 \times 4 + 6 \times 4 = 56$ kbits/sec. The latter approach, however, would "sound better" as more bits are used in the band with the higher energy (a second factor for this improvement is that our ear is less sensitive at higher frequencies.) In the synthesizer, the two decimated and quantized subband signals are interpolated and combined to reconstruct the speech signal. It should be noted that the notions here are subjective, so the only way to verify the system is to actually listen to its output. It should also be pointed out that a practical coding system is not as simple as the one shown above. The speech normally has a very large dynamic range that will cause problems long before the actual quantization noise becomes a problem. This is normally overcome by compressing the signal using μ- or A-law companding. The speech signal is also usually divided into more than two bands and all the bands do not necessarily have the same bandwidth. Since the speech spectrum is actually time varying, an adaptive quantizer is often used in each channel to control the number of bits/sample. In fact, a more sophisticated speech coder can be chosen as a topic for a project.

10.2 Implementation of Multirate Algorithms

Decimation and interpolation systems change the sampling rate, so to implement such a system, the hardware must be able to work at different sampling rates. This is possible with the TMS320C30 EVM. As we have described in Chapter 2 and 3, the EVM board has a TLC32044 analog interface circuit (AIC) that controls the A/D and D/A conversions. Via the transmit register of the TMS320C30, configuration words are sent to the AIC to set the desired sampling rates and operation modes. In the previous experiments, the A/D and D/A (or receive and transmit) sections were set to work in the synchronous mode and the two sections had the same sampling rate. The AIC also has an asynchronous mode where the A/D and D/A sections can work at different sampling rates. It is controlled by the 5th bit of the AIC control register; a one sets the synchronous mode and a zero sets the asynchronous mode. In Chapter 3, we discussed the configuration of the AIC, where we used the hexadecimal word `0x2a7` to set the control register; selecting synchronous mode, primary input, etc. Changing the 5th bit to a zero for asynchronous mode yields a new configuration word: `0x287`. In the asynchronous mode, the receive and transmit sections can be set separately to different sampling rates. As described in Chapter 3, the sampling rates are determined by the values in the A and B registers of the AIC. Each register has two separate parts, which are marked with T and R in Fig. 3.3, to control the transmit and receive sections, respectively. Using different values for T and R allow different sampling rates for transmitting and receiving.

When implementing a system for decimation by an integer factor M, the sampling rate in the receive section is set at the higher rate, F, and in transmit section is set at the lower rate, F/M. The input samples are read in the receive interrupt routine and the decimated output samples are written in the transmit routine. As pointed out in Section 10.1.1, by using FIR filter the lowpass decimation filter can be computed at the lower rate F/M. Thus, the filter computation should be placed into the transmit interrupt routine. To implement an interpolator, the receive section is set at the lower sampling rate F and the transmit section is at the higher rate LF. Similarly the input sample is received in the receive interrupt routine and lowpass filtered in the transmit routine. Before being filtered, however, the input signal must be filled in $L-1$ zero values between each pair of input samples. This need be done carefully. It should be noted that it is possible to do interpolation without really filling in zero-valued samples as implied by Eqn. (10.12). In order to make implementation easier, we can simply fill in zero-samples and use Eqn. (10.10) to compute the filter output. The implementation of a FIR filter has been discussed in Chapter 5. The program developed in that chapter can be directly replanted here.

The implementation of the filter bank system shown in Fig. 10.7 is a little tricky. As discussed in Section 10.1.3, $x[n]$ is sampled at sampling rate F and then divided into two channels with sampling rate $F/2$. The synthesizer interpolates the two decimated channel signals back to the sampling rate F and combine them to produce $x'[n]$. Therefore, the system requires that both the receive and transmit sections must be operating at the sampling rate F. To represent the decimated channel signals, however, we need a sampling rate of $F/2$. A solution is found when we notice that the decimated channel signals do not need to be transmitted, so we can avoid the sampling rate $F/2$. However, parts of the internal system must still

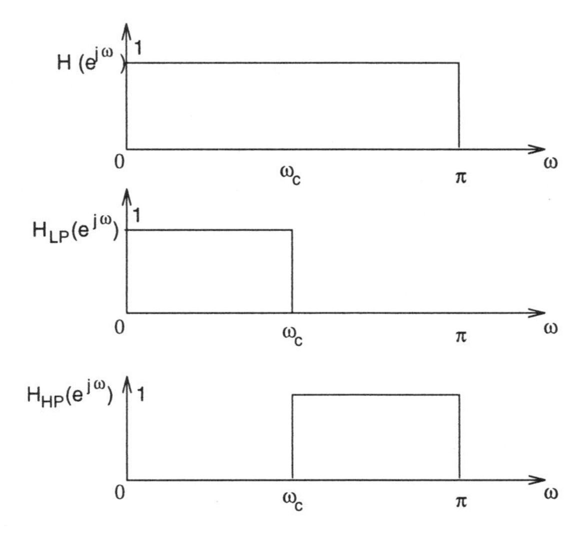

Figure 10.8: Graphical description of a highpass filter calculation

operate at the $F/2$ sampling rate, but this can be achieved by updating only these parts at every other interrupt (sample arrival). Toggling a flag at each interrupt this is easily accomplished using a BR instruction.

In Section 10.1.3, we saw that a lowpass or bandpass filter need be used for decimation or interpolation system, respectively. However, when the system is implemented on an EVM board, we can replace the bandpass filter with a highpass filter because the AIC has an anti-aliasing filter in A/D section and an anti-imaging filter in the D/A section that cut off frequency components above $F/2$.

Using a highpass instead of a bandpass filter allows us to reduce both the storage and computational requirements by a factor of two. As illustrated in Fig. 10.8, a highpass filter can be obtained by subtracting a lowpass response from a constant magnitude (all-pass) response. This results in an impulse response

$$h_{hp}[n] = \delta[n] - h_{lp}[n] \tag{10.13}$$

where $h_{lp}[n]$ represents the lowpass filter coefficients. The delta function in Eqn. (10.13) will only affect the $n = 0$ term of the calculations, which allows us to use the same set of filter coefficients for both the lowpass and highpass filters. Since the two filters operate on the same input signal, they can also share the filter calculations. When computing the highpass filter, we can put the negative sign in Eqn. (10.13) into the accumulation of the products and make a final correction for the $n = 0$ term.

Once the filter bank system is working, we can use it to conduct experiments on speech coding. Using a speech input signal to the system, we can listen to the reconstructed (encoded and decoded) speech output. As discussed previously, we can reduce the number of bits in each channel to see and hear the effect on the output speech. The quantization can be done by masking-off bits using the `AND` or `ANDN` instructions, as we did in the experiments on quantization effects in Chapter 8. Since the filter computation is implemented with floating-point instructions while the speech signal is represented in a fixed-point format, a conversion between floating-point and fixed-point data need be done to the two channel signals before and after the quantization process.

10.2.1 Experiment 10A: Decimation by an Integer Factor

Implement a sampling rate reduction by a factor of 2 on the TMS320C30 EVM. Select an 8 kHz sampling rate for the input and a 4 kHz sampling rate for the output. Use a sinusoidal sweep as the input signal. The experiment can be divided into three steps:

10A.1 Implement a sampling rate compression by a factor of 2 without any prefiltering. Observe the aliasing in the output.

10A.2 Insert a lowpass filter to remove the aliasing components and obtain the correct decimated signal. Record your observations and comment on them.

10A.3 Modify your FIR filter, such that it takes advantage of the fact that only $1/M$ of the filter outputs are "needed."

10.2.2 Experiment 10B: Interpolation by an Integer Factor

Implement a sampling rate increase by a factor of 2. Select a 4 kHz sampling rate for the input and a 8 kHz sampling rate for the output. Use sinusoidal and square wave as input, respectively. Similarly, carry out the experiment in two steps:

10B.1 Implement a sampling rate expander by a factor of 2 without any postfiltering. Observe the imaging effect in the output.

10B.2 Insert a lowpass filter to remove the imaging components and obtain the correct interpolated signal. Record your observations and comment on them.

10.2.3 Experiment 10C: A Filter Bank System Using Integer-band Decimation and Interpolation

Implement the two-channel filter bank system shown in Fig. 10.7. Assume that the input has a 4 kHz bandwidth. Use a sinusoidal signal as input. The experiment can also be carried out in two steps:

10C.1 Implement only the analyzer part from Fig. 10.7 (leave out the quantization) and use the D/A converter to output channel 0 and channel 1, respectively, and observe them on the oscilloscope. Verify that the spectrum of channel 1 is an inverted version.

10C.2 Implement the entire filter bank system in Fig. 10.7. Use the method described in Section 10.2, or any other way you find, to simulate a downsampling by a factor of 2 for the decimation part. Output the interpolated channel 0, channel 1, and the combined final signal, respectively, via the D/A converter. Observe them on the oscilloscope and comment on your system.

10.2.4 Experiment 10D: Subband Speech Coding

10D.1 After you have accomplished Experiment 10C, use the system for speech coding. Connect a tape recorder and a speaker to the EVM board. Use a speech tape as the input and listen to the reconstructed speech output in the speaker. Since the EVM output signal is not powerful enough to drive a speaker, an amplifier is also needed between the EVM output and the speaker. The recorder, speaker, and amplifier should have a decent transfer function in the $0 - 4$ kHz range. Reduce the number of bits in each channel gradually and listen to the effects. Try to use different speech segments and compare the results. Record your findings and comment on the coding system.

10D.2 Using the integer-band decimation and interpolation described above, but break the input speech into 4 bands: $0 - 1$ kHz, $1 - 2$ kHz, $2 - 3$ kHz, and $3 - 4$ kHz. Repeat the coding process in 10D.1. Compare the results with 10D.1 and comment on them.

Chapter 11

DSP Projects

This chapter presents brief project suggestions for a 4 to 5 week project. To get some idea of what might constitute a project, the second section describes some completed projects in more detail.

11.1 Projects Suggestions

Here is a set of project ideas. The descriptions are deliberately brief as we would like for you to use your imagination to define the scope and direction of each project. During this definition stage, you are encouraged to discuss your project selection with your instructor.

Every project *must* be approved by your instructor before you embark on it. To get this approval, you should hand in a one-page description of your project before the designated deadline.

During the last class, there will be a presentation of all projects, where each project group will have 15-20 minutes to describe and demonstrate their achievements.

- A Wigner spectrum analyzer similar to the FFT spectrum analyzer from chapter 7
- A wavelet spectrum analyzer similar to the FFT analyzer from chapter 7
- Serial port RS232 interface
- Speech coder compressing speech according to the G721 (32kbit/s ADPCM) ITU standard
- Speech coder compressing speech according to the G728 (16kbit/s CELP) ITU standard

- Building an interface between Matlab and the TMS320C30
- An FFT segmentation algorithm to overcome the 1k limit implied by the internal memory size
- A simple MIDI sound synthesizer
- A spread-spectrum communication channel between two DSPs
- A 300 (v.21) or 1200 (v.23) baud FSK modem
- An acoustic echo canceller for speaker phone based on multiple sensors (hardware required)
- A voice-over-data communications channel
- An 8kHz to 11kHz sample-rate converter that can change the sampling rate of discrete sequences
- Efficient implementation of pruned FFT algorithms where only part of the spectrum is needed
- Implementation of robust filter structures
- Pitch extraction for speech signals
- Speech recognition system with limited vocabulary (f.ex 10 digits)
- A multi-tone generator and receiver for phone applications
- Spectrum estimation techniques
- Pitch changer (Speech compressor)
- Matlab filter design with automatic code generation
- Speech interpolation/decimation
- Viterbi encoder/decoder for V.34 ITU modem standard
- Scrambler/descrambler for V.34 ITU modem standard
- An IS-54 equalizer for wireless communication

11.2 Projects Examples

This section presents a description of five projects. They are included to give an idea of what a project might look like. The projects can also be used as a template for your own projects or can serve as a basis for further work. The complete source code for all the projects can be found on the disk in the back of the book.

11.2.1 A Spread Spectrum Communication System

by **Nageen Himayat and Dean Thompson**

Spread-Spectrum Communication

A spread spectrum communication system is one in which the bandwidth of the signal is in excess of the minimum necessary to send the information. The band spread of the signal is accomplished by means of a code that is independent of the data. Synchronized reception with the code at the receiver is used for de-spreading and subsequent data recovery.

To understand the basic principles behind the spread spectrum scheme, consider the simple base-band scheme shown in Fig. 11.1. Here, the original message $m(t)$ is a binary sequence that is transmitted by using a pulse of duration T seconds, giving a data rate of $1/T$ bits/second. The power spectrum of the message $M(f)$ is also shown. In the direct-sequence spread spectrum technique a pseudo random (PN) sequence consisting of pulses of alternating polarity (known as a chips), each of duration T_c seconds, is used to modulate the message waveform. The chip duration T_c is much smaller than T and $N = T/T_c$ is the length of the PN sequence, which determines the spread factor. Hence, the effective rate of transmission is now $1/T_c$ chips/second corresponding to a much larger bandwidth than the original message sequence. To give an example, Fig. 11.1b shows a PN sequence of length 7, (1, -1, 1, -1, -1, 1, 1), which is used to spread the message shown in Fig. 11.1a. This results in spreading of the spectrum by a factor of $N = 7$. Fig. 11.2 shows a typical transmitter and receiver of a base-band spread-spectrum system. In lieu of the correlation receiver shown in the system, it is also possible to use a filter matched to the coded signal. When the transmitted signal is in exact alignment with the filter, sharp peaks in the filter outputs are obtained that help in deciding between a "one" or a "zero." The matched filter eliminates the need to acquire the exact phase of the code, which is necessary for the correlator structure. In our implementation, we shall be using a matched filter receiver.

Although the bandwidth required to transmit the modulated message is much larger, there are several advantages of this scheme. As the signal is occupying a much larger bandwidth, it is more immune to narrow-band interference, multipath effects, and malicious jamming with finite power. This scheme also provides for secure communications because the spreading code is available only to the user. Currently spread-spectrum techniques are also being used to provide multiple access capabilities in communication systems. We shall not discuss the details of spread-spectrum systems as several texts provide excellent coverage. Interested readers are referred to [4, 22].

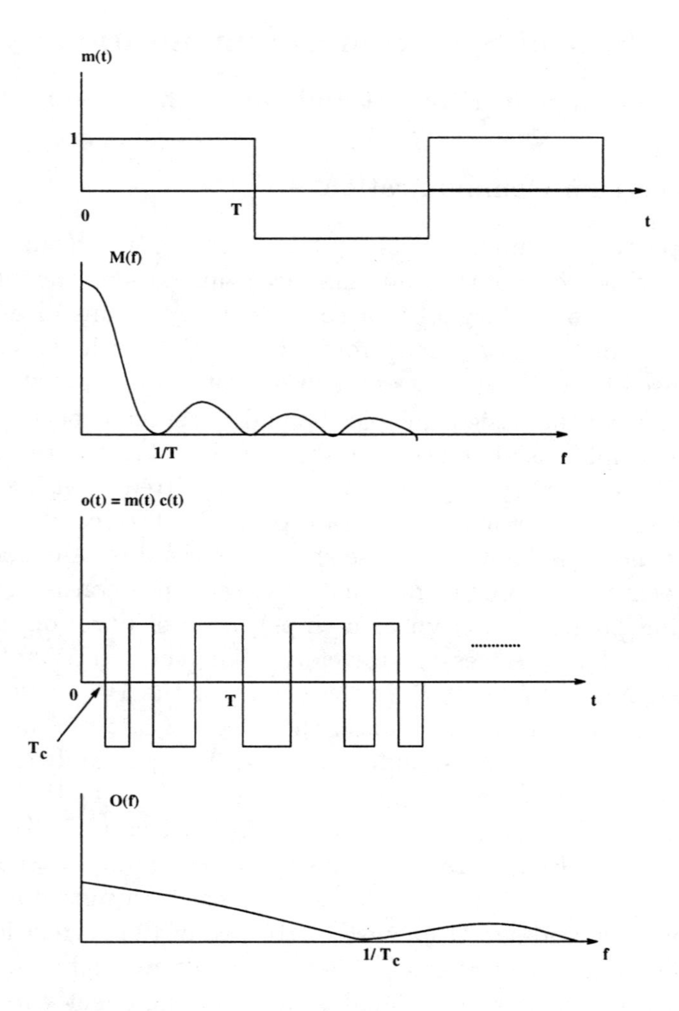

Figure 11.1: (a) Conventional data transmission; (b) Spread-Spectrum data transmission

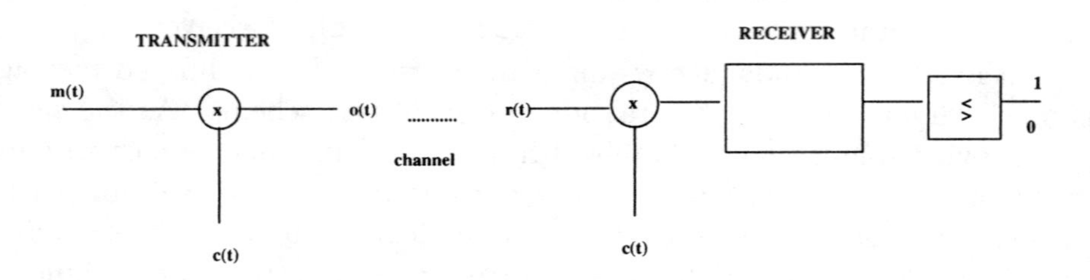

Figure 11.2: Schematic representation of a base-band spread-spectrum communication system

The basic goal of this project is to demonstrate communication between two evaluation modules. Most projects considered so far have focused on implementing algorithms on a single DSP. This project will illustrate some of the problems that arise when two DSPs are required to communicate; even in the simplest base-band scheme synchronization becomes a problem as the clocks of the two DSPs may be slightly off-set from each other. This project also provides an opportunity to investigate the principles behind spread-spectrum techniques. Limitations are imposed on system design by the fact that the maximum sampling rate available is 19.2 KHz. Therefore, either very low data rates need to be used or one has to employ small spread factors. Our motivation, however, is to use this project as a demonstration tool. We envisage that the issues investigated in this project can provide guidelines for projects involving more complex DSP-based communication systems.

Project Implementation

The setup of this project is as follows. The user enters the message via the keyboard of the transmitter-host-PC (which houses the transmitting EVM). Each character is converted into ascii and written to the communication register used for transferring data between the host and the C30 (for details of host-pc and C30 data communications refer to Appendix A). The C30 then reads from the communication register, and forms the appropriate message sequence, which is then modulated by the code sequence. The Analog Interface Controller (AIC) converts the binary stream to an analog waveform. On the second PC, the AIC samples the received signal. The signal is passed through a filter, which is matched to the transmitting code sequence. The receiver then searches for the peaks of the matched filter which it uses to decide whether a "one" or a "zero" was transmitted. Since the transmitting and receiving clocks are slightly off-set, we need to worry about clock synchronization. In this implementation, we make sure that the peaks in the output of the matched filter are spaced at regular intervals. Symbol synchronization is achieved by padding the 8-bit ascii word by 8 "ones." Therefore, 8 "ones" followed by a "zero" signals the start of the ascii word (each ascii word starts with a zero). The received symbol is written to the communication register of the receiver-host-PC, from where it is read, converted to a character, and written to the screen.

The following design parameters were used in our implementation:

Sampling Rate:	8 KHz
Data Rate:	100 bits/second (approximately)

Chip Rate: 7 chips/ bit (1, -1, 1, -1, -1, 1, 1)

These parameters imply that we need to send roughly 80 samples per data bit or 80/7 samples per chip. We decided to use 11 samples per chip, or 77 samples per data bit, which implies a data rate slightly higher than 100 bits/second.

On the PC, the program for reading and writing to C30 are written in turbo C. The structure of both the receiver and transmitter codes is fairly straight-forward. The main sections of the transmitter code are the `nxtdat`, which determines the bit to be transmitted and `delay`, which modulates the sequence with the PN code. When no data is being written on the PC, the transmitter transmits all "ones." The coded analog waveform may be viewed on the oscilloscope by monitoring the output of the EVM30. On the receiver side, the main sections are the matched filter and the code for peak detection and peak alignment. The matched filter output is expected to peak after every 77 samples. Due to the difference in clock rates between the transmitter and the receiver, the peaks may not occur after 77 samples. Hence, we translate the location of the peak to be in the middle of 77 samples and when this location changes, it is forced back to the desired location by altering the number of samples collected for every bit. The output of the matched filter may be monitored on the oscilloscope. When no data is being transmitted, one observes only positive peaks as only "ones" are being transmitted. The routine `datchk` checks whether a valid data word is received before it is written to the PC communication register. One needs to type the data words slowly on the transmitting PC otherwise C30 will miss the characters that were typed while it was processing the previous character.

Demonstrations

This project may be used for investigating various aspects of a spread spectrum communication system.

- The noise immunity of the system may easily be checked. Attach a noise generator to the input of the transmitting EVM30. Add program code in the interrupt service routine to read the noise samples from the input and to add the values to the output samples of the transmitted waveform. Change the noise level until you observe a lot of transmission errors.

- We can check the security aspects of a spread spectrum communication scheme by simply changing the transmitting PN code. As the receiver filter is no longer matched to the transmitting code, the data can no longer be recovered accurately.

- A spread spectrum system is effective against multipath effects and narrow-band interference. Such degradation can be introduced easily by processing the transmitted signal appropriately. Generate an interfering sinusoidal signal and add it to the transmitted waveform. Observe the effect at the receiver. Similarly, a multipath channel can be modeled by a suitable FIR filter (introduce suitable delays and attenuation in the impulse response). Compare it with a system without spread-spectrum capabilities (change the transmitter and the receiver PN code to all "ones").

- Write a subroutine that will implement the generation of the PN code via shift registers [4, 22]

11.2.2 Matlab Interface to the EVM Board

by **Joe Palovick**

Matlab interface

Matlab is a commercially available numerical analysis program that is available from Mathworks, Inc. This software package allows users to do elaborate mathematical calculations with little or no programming. The TMS320C30 EVM board (EVM) plugs into the backplane of an IBM-PC or compatible (PC) and is a fairly cost-effective method to achieve up to 50 MFLOPS of computing power. The following explanation outlines the operation of an interface between Matlab and the EVM on an IBM-PC. The following list contains operations that are possible with this interface.

- The EVM as a co-processor to Matlab (e.g., floating point data is loaded into the EVM, the FFT of the data is computed, by a program running on the EVM, then the processed data is returned to Matlab).

- To use the A/D converter of the EVM to "record" information into Matlab, so that it could be processed and later "played" on the D/A converter of the EVM.

The interface between Matlab and the EVM board is implemented via extended functions in Matlab, which are written in the C programming language, and a program written in both C and TMS320C30 assembly language on the EVM. The software protocol for this interface involves the PC sending a command to the EVM and the EVM acting on that command in a pre-determined manner. The complications when transferring data across this interface is caused by the fact that the EVM board supports data that is 32 bits in size, while the bus of the EVM's host interface is 16 bits wide. Also, Matlab and the PC support the IEEE floating point format while the TMS320C30 uses an internal floating point format. The cases above require that data be formatted before it can be sent in either direction. The discussion that follows will explain the method in which these data types are transferred, and the protocol that handles the command and data transfer. The procedure for converting the TMS320C30 floating point format to the IEEE format is described in [19] in section 11.3.7 and will not be described here. It should be noted that data transfers wider than 16 bits must be disassembled before transfer and reassembled upon reception. For the cases of integers (32 bits) and floating point numbers (IEEE or C30), two transfers of 16 bits each are necessary. Below is a C code segment showing how floating point data would be transferred from the PC to the EVM. The code below would reside on the PC. The data type "short" is a sixteen bit integer, float is a 32-bit IEEE floating point number, and the function send_short writes 16-bit quantities to the EVM's data transfer register. The conversion from IEEE floating point to C30 floating point would be done on the EVM because it runs much faster than the PC.

```
...
float       floatvar = 1.2;          /* Desired Transfer Value      */
short       *datptr;                 /* Transfer pointer            */

datptr = ( short *)&floatvar;        /* Cast to unsigned short ptr */
send_short(datptr[0]);               /* Send first 16 bits         */
send_short(datptr[1]);               /* Send the last 16 bits      */
...
```

The above code segment illustrates what must be done to transfer data across the backplane of the PC to the EVM. The implementation on the EVM side is slightly different because the EVM must monitor interrupts to determine when the PC is transferring data. As described above, the PC sets different interrupt lines on the TMS320C30 when it is reading or writing data. When the PC writes data to the EVM, interrupt level 1 will be set (See the EVM manual [15]). An example of the PC writing data to the EVM is demonstrated by the following code segment (`ar0` is TMS320C30 address register 0 and points to the command/data register. `r0` is the TMS320C30 extended precision register 0, and is the data value. Also, `if` is the TMS320C30's interrupt flag register.) The subroutine `getint` is called to get a 32-bit value that is returned in `r0`. The code to transfer data from the EVM to the PC was done in TMS320C30 assembly language.

```
getint:   tstb 1,if       ; Wait for PC to write data
          bz getint

          ldi *ar0,r1     ; Get the first 16 bits
          lsh 16,r1       ; Shift it left 16 places (2 MSB's)

getnxt:   tstb 1,if       ; Wait for next host write
          bz   getnxt

          ldi  *ar0,r0    ; Get the data
          and -1,r0       ; Mask off upper bits

          or r1,r0        ; Combine with upper 16 bits

          rets            ; All done
```

The returned value `r0` could be cast to a float or an integer depending on the type of data transfer. If the transfer was a floating point number, then an additional conversion is required to get the data in C30 floating point format. The protocol for transferring data is tightly coupled between the PC and the EVM. The description below will list several functions that exist on the EVM and the method in which the functions are called. In this implementation, the EVM board is always a slave to the PC. The classes of functions that exist on the EVM board are data transfer functions, data manipulation functions, "record and play" functions. The end of this section explains how a typical command sequence is handled between the EVM and PC. The most basic class of functions supported by the EVM are data transfer functions. Transfers are relative to the PC with the functions "getting" or "sending" a specified number of floating point numbers between the PC and EVM. Data manipulation functions, such as an FIR filter, could be implemented by transferring filter coefficients and data points to the EVM and then having the EVM return the filtered data to the PC when the PC requests the data. Functions for "recording" from the A/D converter and "playing" data to the D/A converter will transfer a requested block size of information in either direction doing the conversion to floating point numbers on the EVM board. A typical command sequence between the PC and EVM involves the following steps:

1. The PC sends a command to the EVM board.

2. The EVM acknowledges the command.

3. The EVM performs the command that the PC asks for (either data transfer or executing a function).

4. The PC writes a zero to the command register.

5. The EVM writes a zero to the command register and waits for the next command from the PC.

Steps four and five are present to allow for synchronization between the programs running on the EVM and on the PC. (They would not be present if there were separate registers for data and commands on the EVM. See [15] for an explanation of the command/data register). Implementing the above interface into Matlab is a straight-forward process. The interface breaks down into a data structure in Matlab being passed to extended functions, called MEX files, which are written in C and callable by Matlab. The MEX files incorporate the above interface and allow data transfer between Matlab and the EVM board. The list below is from the Pro-Matlab User's guide section 7.3 and is the recommended protocol for implementing MEX files.

- Validate inputs: check number, type, and size of inputs.
- Allocate outputs: given the sizes of the input matrices, create the output matrix structures and equate them to the appropriate elements of plhs.
- Allocate temporary work arrays.
- Perform desired calculations.
- Return control to Matlab.

Incorporating a PC program to a MEX file is generally a smooth transition because Matlab includes functions to create matrices, report errors to Matlab, etc.

11.2.3 SERCOM. A Serial Voice Communications Link for the TMS320C30

by **Mike Meixler and Anestis Karasaridis**

The objective of this project was to implement a system for real time voice communication between two PCs equipped with TMS320C30 EVMs via an RS232 serial channel. The system is appropriately named SERCOM. SERCOM has an 8-bit dynamic range and requires a 9600 baud RS232 channel. Figure 11.3 illustrates a block diagram of the system.

During operation, SERCOM users, stationed at two PCs connected by an RS232 cable, communicate with each other by speaking into a microphone and listening to a speaker. Like CB radios, the system is half-duplex.

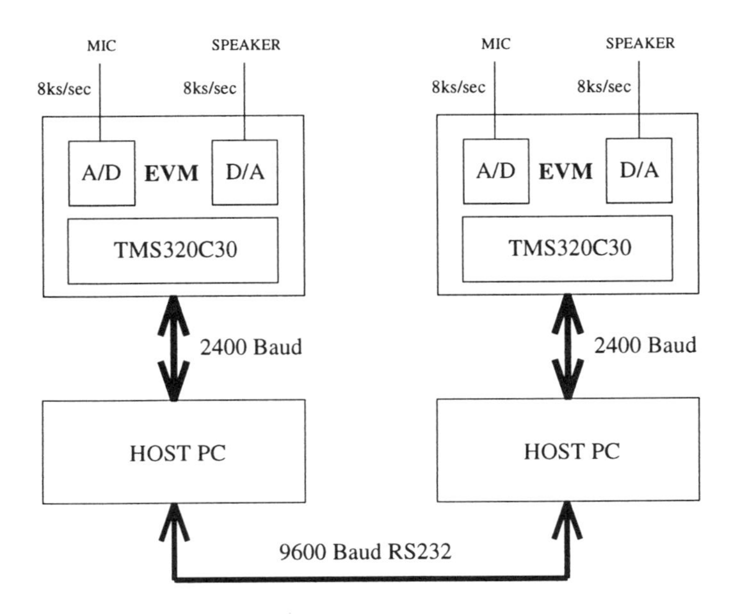

Figure 11.3: Block Diagram of SERCOM System

While one station is transmitting, the opposite station must receive. The users use the keys on the keyboard of the PC to signal the start and end of transmission.

The motivation for implementing this system is a practical one. If it is shown that voice communication can be achieved through a 9600 baud serial channel, one can imagine that the 9600 baud channel could easily be replaced by a long distance digital channel such as Internet, Prodigy, etc. Such a system may render a less expensive alternative to the current inflated costs of telephone long distance in America.

To use SERCOM, load all of the necessary files from the companion disks onto the PCs, following the instructions in the `readthis.txt` files. The COM1 ports of the two PCs must be connected by an RS232 cable.

Theory of Operation

Refer to Fig. 11.4 for a block diagram of one SERCOM station. The system samples voice at the rate of 8 ks/sec and transmits coded data across the RS232 connection at a rate of 9600 baud. With each sample being 8-bits, this implies that a compression ratio of at least 6.67 is needed. DSPs are used to perform this compression in the transmitter and to synthesize the voice signal in the transmitter. A Linear Predictive Coding (LPC) algorithm is used to code and decode the data.

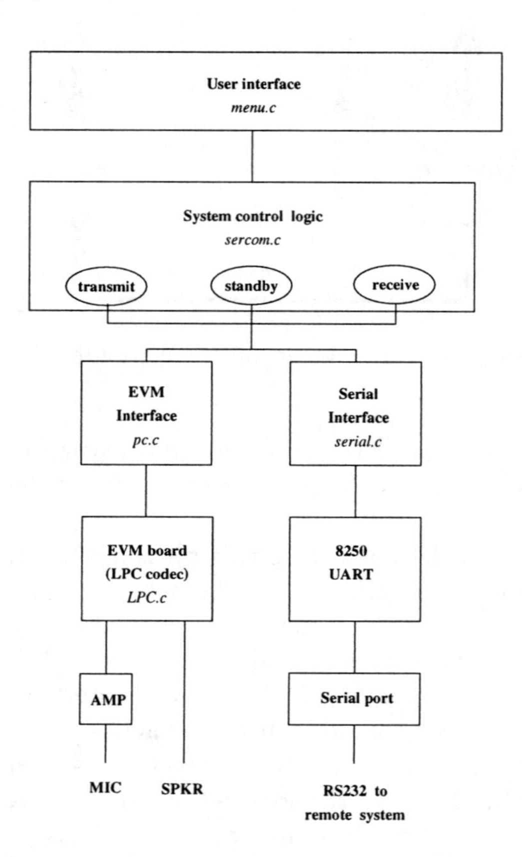

Figure 11.4: Modular Structure of Individual Station

During transmission, the DSP is used to sample the A/D converter connected to the microphone and compress the data to 2400 baud. This 2400 baud coded data is transferred to the memory of the PC through the EISA bus. Control bytes are appended to the packets of coded data. The entire package is sent across the 9600 baud serial channel to the receiver, which performs the inverse operations to synthesize the voice signal.

It is important to note that real time communication is possible only because of the parallel processing between the DSP and the PC. In the transmitter, while the DSP is sampling the voice signal and generating the compressed code, the PC is transmitting the packet of code bytes previously generated. Inversely, in the receiver, while the DSP is constructing the voice signal from previously received code, the PC is receiving the next packet of information.

The implementation of the system software can be subdivided into three subprograms. They are: 1) the DSP subprogram, which handles the sampling, construction, coding, and transfer of data between the DSP and the memory of the PC; 2) the serial communications subprogram, which facilitates the transfer of data between the two stations via the RS232 channel, and; 3) the control logic subprogram, which provides an interface for the user and maintains control of the process. The control logic subprogram runs over the DSP serial communications subprograms and calls subroutines in these subprograms. See Fig. 11.4 for the modular structure of the individual station. All source code for the system is on the companion disk.

DSP Subprogram

The purpose of the DSP subprogram is to transfer commands and data between the PC and the DSP. The DSP subprogram was built from routines borrowed from the Texas Instruments (TI) LPC coder for the TMS320C30 called "LPC". These subroutines interface with the program lpc.out running on the DSP. They are found in `pc.c`. Some of them are slightly modified from their original form in TI's "LPC" program. In transmit mode, these subroutines instruct `lpc.out` to sample the A/D converter, compress the data, and transfer the coded data to the memory of the PC. In receive mode, these subroutines transfer coded data to the memory of the DSP, instruct `lpc.out` to reconstruct the voice signal, and send samples to the D/A converter. Data is transferred between the memory of the PC and the memory of the DSP in packets of six bytes. For an overview of LPC coding and the TI implementation for the TMS320C30 DSP, see the Section "LPC vocoder: An overview."

Serial Communications Subprogram

Subroutines that handle serial communications via the RS232 channel are found in `serial.c`. As is shown shortly, these subroutines must transfer data at the effective rate of 350 bytes/sec. Since RS232 is an asynchronous serial communication channel, there may be idle states between bytes being transferred. Therefore, the actual throughput through the serial channel may be far less than the baud rate that the port is setup for (in this case 9600 baud).

In any asynchronous communication implementation, a system of handshaking must be established. Handshaking is used to prevent the sender from sending information when the receiver is not ready to receive. Thus, in a handshaking system, the receiver must signal that sender when it is ready to receive the next byte of information, i.e., after it is finished processing the previously received byte. Between the time that a character has been sent and the time that the receiver asserts a request for a subsequent byte (i.e., while the character is being processed), the transmission line is idle. For this reason, software efficiency (and machine speed) is essential for fast data transfer. SERCOM was developed on a 286 PC running at 10 MHz.

The minimum required throughput of the RS232 channel of 350 bytes/sec is determined as follows: The LPC coder produces 2400 baud code in packets of three 2-byte words. This corresponds to 6 byte packets generated in 20 ms increments. The system appends a seventh control byte onto this packet (discussed later), so the required throughput of the RS232 channel is 7 bytes/20 ms, or 350 bytes per second. Figure 11.5 shows the communication protocol used.

Hardware handshaking, rather than software handshaking is used in order to meet the speed requirement. The handshaking method used utilizes the RTS (request to send) and the CTS (clear to send) lines in the serial cable to indicate when the receiver is ready to receive a subsequent byte. The serial cable must be a null-modem type that criss-crosses the RTS and CTS lines of the two serial ports. A timing diagram for handshaking sequence is shown in Fig. 11.6. In this implementation, the receiver first asserts RTS to indicate that it is ready to receive a byte. When the transmitter detects this low-to-high transition by polling the CTS line, it sends the byte. After receiving the byte, the receiver returns the RTS line to low, then proceeds to process the byte. Once ready to receive the next byte, the receiver asserts RTS again, and the process repeats. A timing diagram for the handshaking scheme is shown in Fig. 11.6.

From a programming standpoint, all of the RS232 operations discussed above including the polling and asserting of the handshaking lines and the

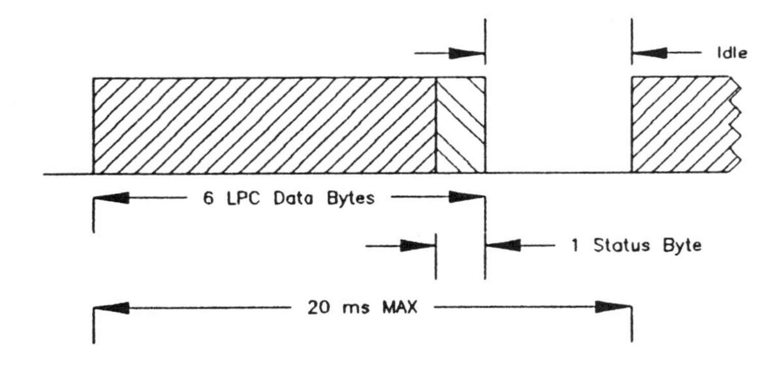

Figure 11.5: Communications Protocol

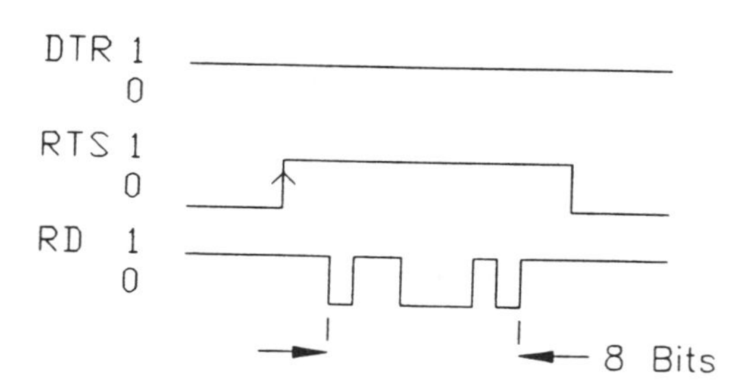

Figure 11.6: RS232 Handshaking Timing Diagram

sending and receiving of data, are handled through the UART for the COM1 port. The registers of the UART reside in the memory of the PC at locations `0x3f8` through `0x400`. These addresses are written to, or read from, during communication. All of the subroutines used for serial communications are in the `serial.c` file.

Control Logic Subprogram

The file `sercom.c` contains subroutines to handle the control logic of the system. During run time, program control at each of the two stations passes to any one of three states. These states are transmit, receive, and standby. From a user's standpoint, this means that the station can either be transmitting voice, receiving voice, or standing by, i.e., waiting for an incoming "call," or waiting for the user to initiate a call. The standby state in SERCOM is analogous to the telephone on-hook state in the phone system. From a programming standpoint, these states are loops that run continuously until an event occurs, causing exit into another state. Each of these loops are within a subroutine appropriately named after the corresponding state. Refer to Fig. 11.7 for a state diagram of the system and see the companion disk for a commented program listing.

Upon running SERCOM, the program defaults to standby. The standby state has two substates, i.e., remote station ready and remote station not ready. The Boolean variable `rssb` (remote station standing-by) in the standby subroutine is used to specify whether the remote station is also standing by. If it is, then `rssb` is set to 1. Upon start-up, `rssb=0`. Once communication has been established, `rssb=1`.

Once communication has been established and both stations are standing by, control characters are exchanged between the two stations to indicate which state each is in. An 's' indicates standby. When the user of one of the stations initiates a transmission (by pressing the F1 key), the control character 't' is sent to indicate the pending transmission. If a control character is not received within several (10) polls, the program returns to the remote station not-standing-by substate.

Upon entering the transmit mode, the program first displays the appropriate menu and then sends the transmit control character, 't', to the remote station. This control character signals the remote station to enter receive mode. From this point forward, the transmitting station assumes the server role, while the receiving station assumes the client role.

Upon receiving the 't' control character from the remote station, SERCOM transfers states from standby mode to receive mode. Upon entering receive mode, SERCOM displays the appropriate menu.

During transmission, both programs loop continuously in synchroniza-

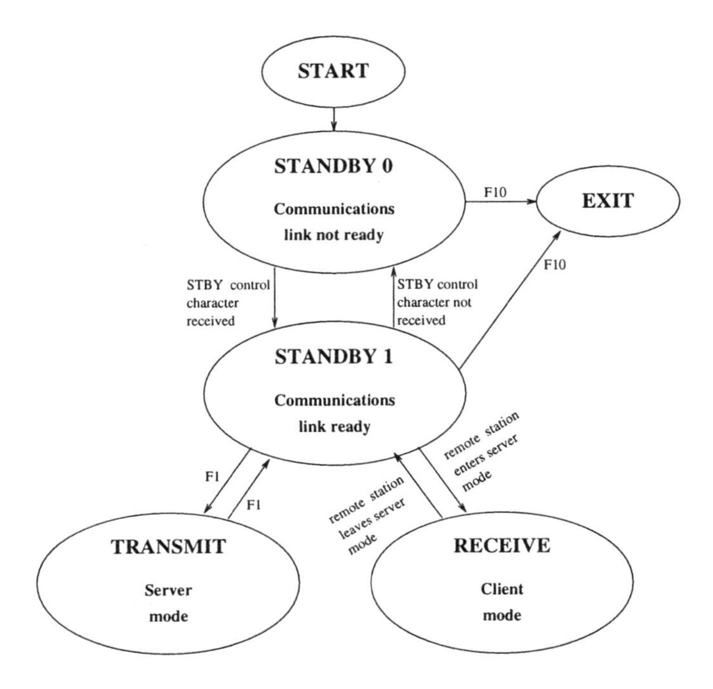

Figure 11.7: State Diagram of System

tion. The transmitter repeatedly fetches data from the A/D converter, compresses it, and sends it across the RS232 channel. Simultaneously, the receiver repeatedly fetches data from the RS232 channel, synthesizes the voice signal from the data, and sends the samples to the D/A converter.

This process continues until the user of the transmitter hits the F1 key to end transmission. At this point, the transmitter appends an 's' control character to the last packet of data transmitted to instruct the receiver to return to standby mode. Next, both stations are in standby mode and either can initiate a transmission. At this point, the receiver may respond to the message just received.

In this manner, a conversation may continue indefinitely. The process finally ends when one party presses F10 to leave SERCOM, thereby exiting standby mode.

The LPC Vocoder: An Overview

LPC is a coder that is based on a linear predictive scheme. This coding method was basically designed for voice signals for low bandwidth specifications. LPC falls into the category of parametric coding methods. This means that instead of a coded version of the quantized input signal, only

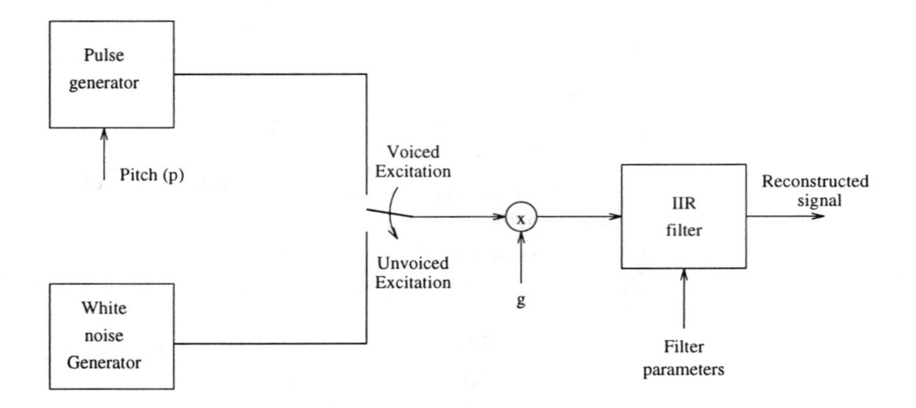

Figure 11.8: Reconstruction of the Voice Signal on the Decoder Side

some parameters of the signal are being transmitted. The parameters in this case are used to determine the coefficients of an IIR filter, which stands as the linear predictor for signal reconstruction. The reconstruction process also involves a signal used as excitation of the IIR filter. This signal is either a white noise or a pulse train with variable period. Figure 11.8 outlines the reconstruction phase of the LPC algorithm.

As it was pointed out above, LPC vocoders are usually applied when there are stiff bandwidth requirements. So they are actually very efficient. In the TI implementation, on which our Serial Communication project is based, an 8kHz sampled voice signal with two words per sample can be coded with 2400 bps. This corresponds to 53:1 compression ratio. There are also LPC vocoders that can compress the signal to as low as 800 bps. These huge compression ratios are achievable by using a large variety of heuristic algorithms. The reconstructed voice signal is highly intelligible, but suffers from spectrum distortion which makes the identification of the speaker difficult.

Conclusion

SERCOM was successful in achieving what it was designed for. We showed that intelligible voice can be transferred digitally through a 9600 baud serial channel. The major limitation of the system is the poor voice quality.

As implemented, the system has little practical functionality because the distance between the two stations is limited by the length of the RS232 cable (< 100 feet). However, the project did serve to show that the available technology can be used to communicate by voice through a 9600 baud channel. A significant improvement to the system would be the modification

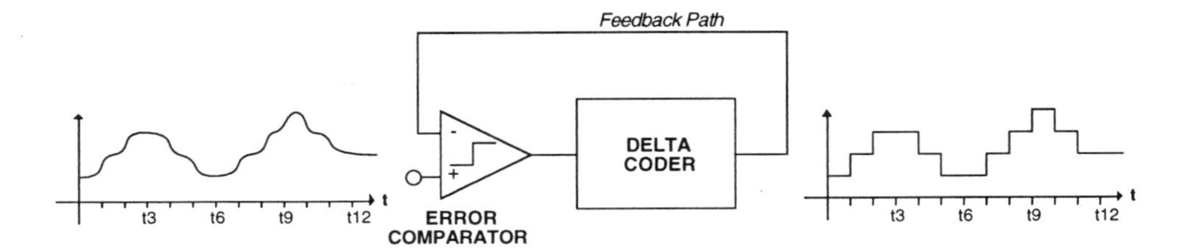

Figure 11.9: A simple delta modulator

to interface SERCOM to a wide area network (bandwidth > 9600 baud). This would enable people separated by very large distances to communicate through SERCOM. Such a system is likely to be more cost effective than the current long distance telephone systems in America.

11.2.4 Continously Variable Slope Delta Modulation

by **Daniel Charnitsky and Philip P. Trahanas**

Continuously Variable Slope Delta Modulation (CVSDM) is an audio encoding technique that is frequently used in military communications equipment. The coding scheme converts an analog voice signal into the digital world, achieving data compression in the process. The CVSDM demodulator is unique in that it can begin decoding the bit stream starting at any arbitrary location. This lack of digital word framing results in very simple synchronization protocols.

Traditional Delta Modulation

Delta modulators operate on the principle of signal error tracking. An analog signal is continuously compared with its reconstructed digital version, as shown in Fig 11.9. If the reconstructed version is lower in amplitude than the original, the ouput of the codec is increased by a fixed amount. Likewise, if the reconstructed signal is larger than the input, the codec decreases its ouput with respect to the previous time slot.

If the clock rate and comparator step size are chosen appropriately, the decoder output will be a fairly good staircase approximation of the input signal. If the step size is too small, the delta modulator will not be able to "keep up" with the steepest slope of the input signal, resulting is slope-overload distortion. If the selected step size is too large, however, the decoder output will "hunt" when the input signal changes slowly. This distortion is sometimes called granular noise or hunting noise. Both forms of distortion are depicted in Fig. 11.10. With a simple delta modulator, the

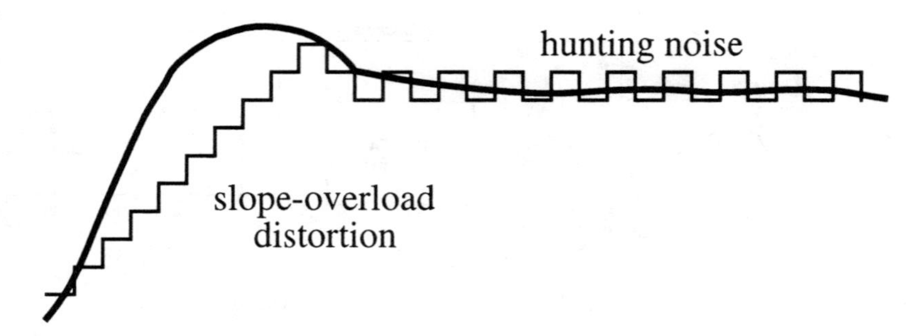

Figure 11.10: Slope-overload distortion and hunting noise

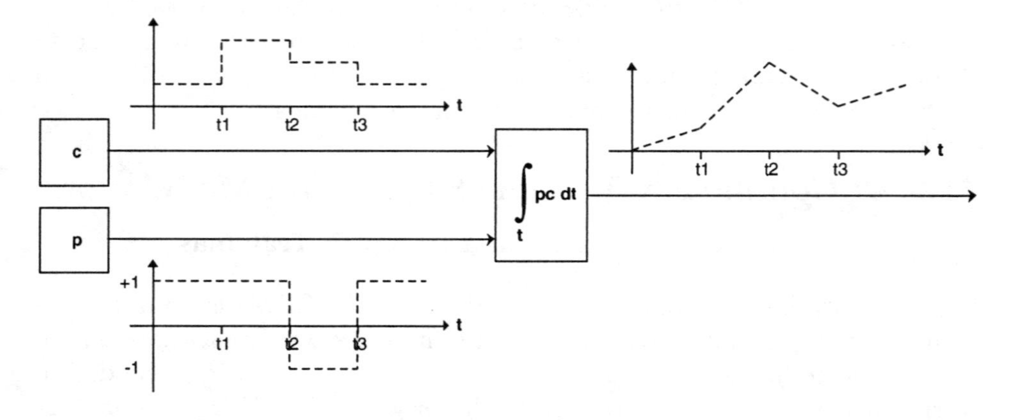

Figure 11.11: How polarity and slope create a desired waveform

step size is fixed to minimize either the hunting noise or the slope-overload distortion.

Continously Variable Slope Delta Modulation addresses this problem, as the name suggests, by adding the ability to dynamically adjust the step size during encoding and decoding. As a result, CVSDM can adapt itself to minimize either the slope-overload distortion or the hunting noise, depending upon the local characteristics of the input signal.

Continously Variable Slope Delta Modulation

The basic principle behind CVSDM is shown in Fig. 11.11. Suppose we are able to create two signals, c and p, which when applied to the integrator, produce a desired waveform. The signal p acts as a polarity control. Notice that the signal takes one of only two values, ± 1. When p is positive, the integrator integrates in an upward direction. Conversely, the integrator integrates in a negative-going fashion when p is negative.

The signal c corresponds to the integrating slope of the integrator. Be-

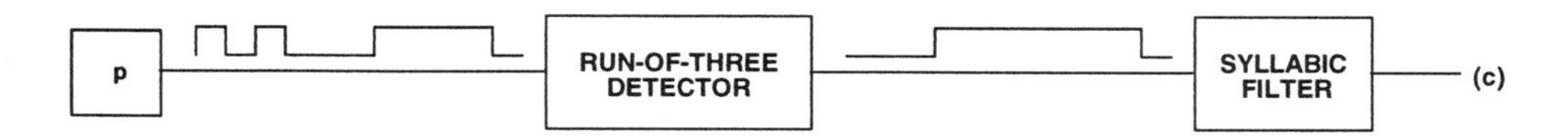

Figure 11.12: How the polarity is used to generate the slope signal

cause it represents the amount the output will increase or decrease over a single clock cycle, it is analogous to the "step size." If c is fixed at a constant value, we have a simple delta modulator. In CVSDM, the signal c is changed based upon the past "performance" of the codec.

The signal c is actually derived from the polarity signal and is not a completely independent quantity as Fig. 11.11 would imply. Fig. 11.12 shows that c is constructed from the polarity using a run-of-three detection algorithm. Whenever a continous run of three or more logical zeros or ones (where logical zero corresponds to -1 and logical one corresponds to $+1$) is present, the run-of-three output is forced logically high, otherwise it is low. The output of the run-of-three detector is applied to a low-pass filter termed the "syllabic filter" to smooth the result and cause the signal to vary slowly with respect to the audio signal that is being encoded.

The relationship amongst the time constants of the syllabic filter, polarity-controlled integrator, and the master sampling clock is key to CVSDM operation. The syllabic filter has a significantly slower time constant than the other two. Loosely speaking, it is desirable for the syllabic filter output to appear as a constant (DC) value to the integrator. In this manner, the filter is geared to track speech energy as a broad quantity. The master sampling clock and the integrator will be responsible for reproducing the fine details of the audio signal.

Anti-Aliasing Filter

An interesting feature of CVSDM is that there are few differences between the encoder and the decoder. Once one design is completed, the other is nearly finished as well. The major changes required for the decoder are:

- There is no need for the comparator or the feedback path in the decoder.
- An anti-aliasing filter is added to the integrator output.

The output filter is needed to remove the high frequency "jaggies" present in the reconstructed integrator output.

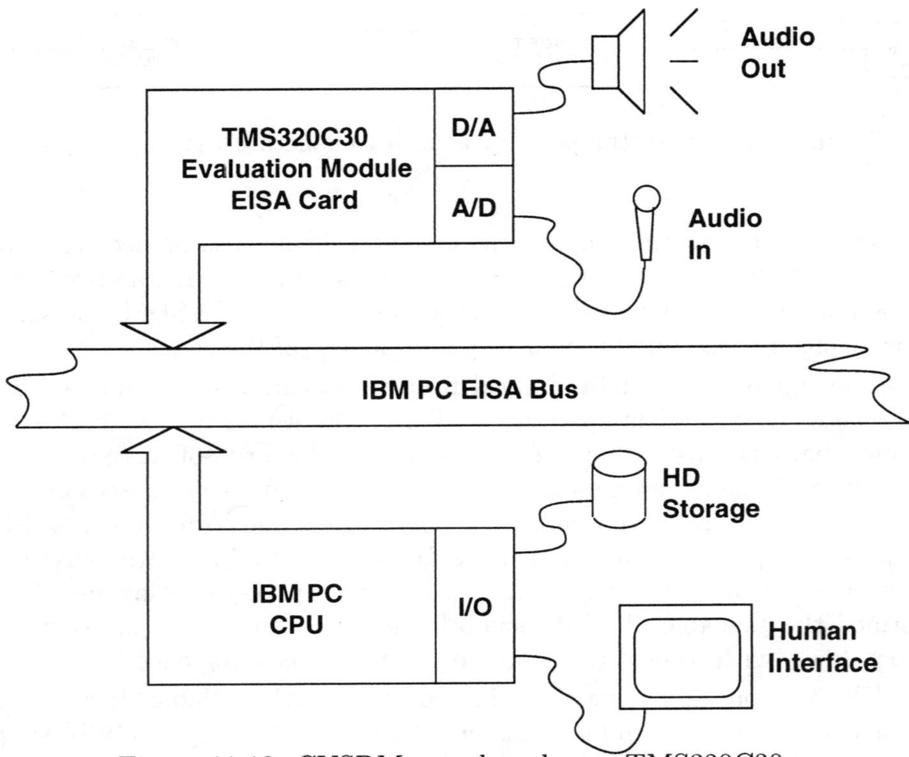

Figure 11.13: CVSDM speech codec on TMS320C30

Project Implementation

The CVSDM codec was implemented on a TI TMS320C30 Evaluation Module (EVM) and a host IBM-PC. The development system provides the user with two modes of operation:

REC Record analog voice, encode the data using CVSDM in real-time, and store the encoded data in a file on the PCs hard disk.

PLAY Retrieve CVSDM encoded data from a file on the PCs hard disk, decode the data in real-time, and play the reconstructed voice through a speaker.

Encoder and Decoder Software

The CVSDM encoding and decoding algorithms are implemented by two independent software modules. The appropriate routine is loaded into the EVM by the host PC in response to a user's mode selection.

When in RECORD mode, the DSP samples the analog input and performs an A/D conversion at 19.2 kHz, the highest sample rate avaliable on the EVM. The sample is then processed using the CVSDM algorithm. The output of the encoder is the sign bit, which is applied to the polarity-controlled integrator. This single bit is packed with previous output bits into a 16-bit word and transferred to the host PC for storage on the computer's hard disk. Packing the bit stream into a word allows for more efficient utilization of the EVM-PC interface and provides more compact storage.

When in PLAY mode, the host PC transfers a 16-bit word to the EVM. The DSP unpacks the word and applies a single bit to the decoder at each clock cycle. The decoder is nearly identical to the encoder; the only difference is the addition of a low pass filter at the output. The low pass filter reduces the high frequency ripple caused by the delta modulation. The filter is a 17th order Butterworth implemented in cascade form.

Host PC Software

The host software is a simple menu driven DOS application for loading and controlling the CVSDM algorithm running in the EVM. The application also provides a convenient method of storing and retrieving encoded data for analysis and demonstration.

Data is transferred to and from the PC via the 16-bit communications port on the EVM board. This port is directly mapped into the PC I/O address space and is easily accessed using the Turbo C functions inport() and output().

The data transfer is synchronized on the EVM by an event interrupt that occurs whenever the host PC accesses the communications port. Synchronization on the host side is accomplished by polling the EVM's READ/WRITE acknowledge status register.

11.2.5 A 300 bps Modem

by **James Tau and Ximing Chen**

Digital Signal Processor(DSP) chips are specialized processors that facilitate the computation of certain operations essential to discrete time signal processing. DSPs are increasingly finding occassions as dedicated processors in addition to general purpose microprocessors. For example, such computationally intensive frills as sound processing and real time graphics manipulation can be dished out to a DSP that frees up the central processor for other more important tasks. Another possible use in this vein is a voice-band modem whose implementation is the subject of this section.

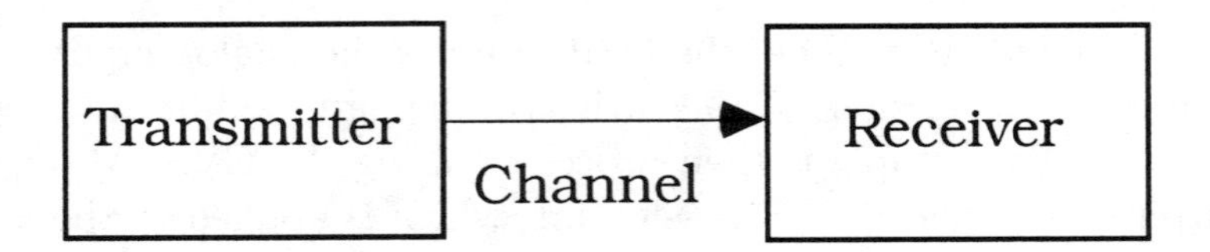

Figure 11.14: Basic communication model

Introduction

Transferring digital data over voice-band channels requires the use of carrier frequencies that are within the bandwidth of the channel to represent the ubiquitous ones and zero. Perhaps the simplest implementation of a modulation–demodulation(modem) device is that specified in the CCITT(now ITU-T) V.21 standard [1], which is the first full-duplex modem to operate over the General Switched Telephone Network (GSTN). In our project, we base our implementation of a 300 bps modem roughly on this standard. The backbone of the V.21 standard is its modulation scheme, which employs binary frequency shift keying and uses frequencies at 980 Hz(1) and 1180 Hz(0) for channel 1, and 1650 Hz and 1850 Hz for channel 2. Our implementation neglects the second channel, which obviously makes it half-duplex. The demodulation method used in our implementation is non-coherent demodulation, which is the most straight-forward demodulation technique of all, though its utility decreases as the signal constellation increases.

Components

The setup for our modem is simple. One computer is designated as the transmitter and another is the receiver. The transmitter runs the modulation program and the receiver runs the demodulation program. The modulation program interfaces with the host computer to receive character input from the keyboard whose ascii code is modulated accordingly. At the receiver, the demodulation program interfaces with the host computer, which displays the received character on the screen. Figure 11.14 shows the basic communication model.

The Transmitter

The heart of the transmitter is the continuous phase binary frequency shift keying(CPBFSK) modulator. Characters are input from the keyboard and are individually transferred to the C30 from the host PC via a front-end interface. A block diagram of the transmitter is shown in Fig. 11.15. Each

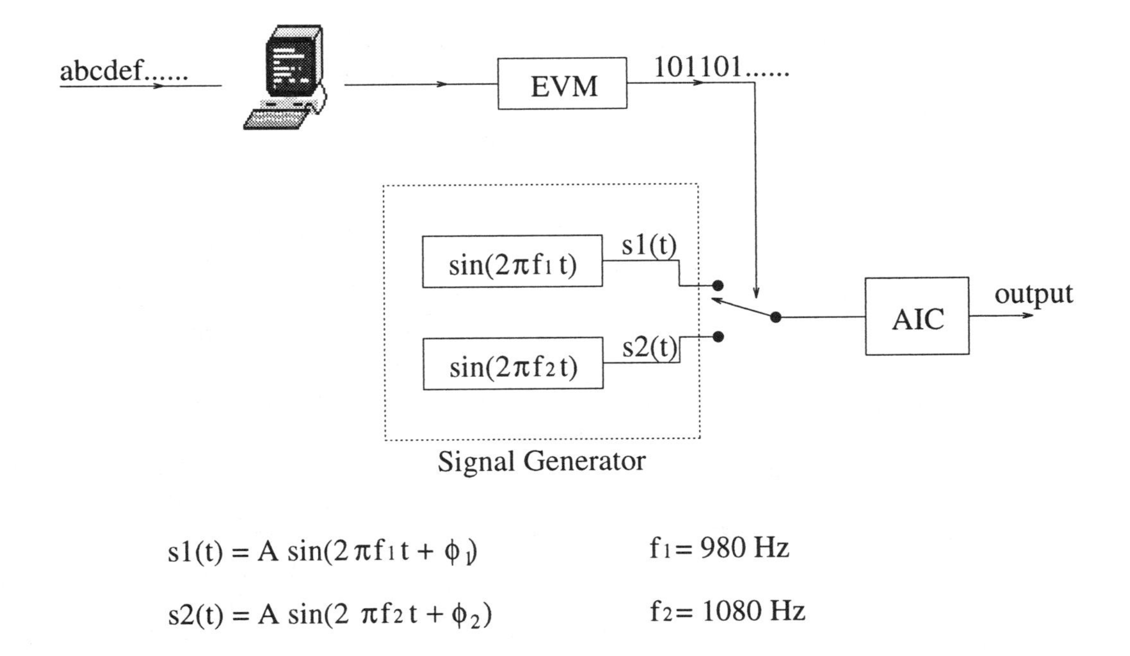

Figure 11.15: Transmitter

bit of the input character is translated into either a high(1180 Hz) or a low(980 Hz) frequency depending on whether it is a 0 or an 1.

Given that the desired symbol rate is 300 Hz, and the DSP sampling rate 8000 Hz, it follows that each information bit consists of a sequence of 27 samples. The sinusoidal waveform generation in our modulator is not produced through the use of a second order difference equation like that used in an earlier experiment but rather by means of a simple lookup table. The major reason behind doing this is efficiency. Both the magnitude and the trignometic value are obtained in one step. At the same time, phase adjustment between successive wave trains becomes much easier to deal with than otherwise. Phase adjustment between successive wave trains is desirable because discontinuous signals introduce unwanted high frequency components that disperse signal power. The same trignometric table is also used in the demodulator as the basis functions by virtue of its availability.

In general, there are two transmission modes for data communication, namely, synchronous and asynchronous transmissions. In asynchronous transmission, each input character is prefixed with a sequence of start bits, followed by a series of stop bits. This is shown in Fig. 11.16. The start and stop bits are used by the demodulator to synchronize symbol timing so that information bits can be properly detected. Synchronous transmis-

Send: ABCDE.	Receive: ABCDE
start-A-stop ... start-E-stop	start-A-B-C-D-E-stop

Figure 11.16: Transmission modes

sion is characterized by having a group of characters enclosed within the start and stop bits, see Fig. 11.16. Usually, asynchronous transmission is used during interactive communication, while synchronous transmission is used for file transfers. As mentioned earlier, we will only implement asynchronous transmission. For synchronous transmission, it merely requires the placement of successive data bytes in a buffer and have `bfsk` steping through them, attaching start bits only for the very first byte. At the same time, the correlator must be modified so that after the receipt of the start bits, it blindly interprets subsequent bytes as ASCII codes. Without impinging on the operation of the modem, we will place only start bits before our character bits and will not add stop bits after them. We choose our start sequence to be a 9-bit code, 101010101, which has a decimal value of 341 that does not correspond to any 8-bit ASCII character. At the receiver, every detection of this special eight-bit sequence indicates that the next 8-bits represent a character.

The Receiver

The workhorse behind the receiver is the correlator, which takes a sequence of input samples belonging to one symbol sampling period(300 Hz) and computes its correlation value. It does this once every sample. Fig. 11.17 shows the receiver set up in detail. The correlation is calculated according to the following equation:

$$r_{sin} = \int_0^T x(t)\sin(2\pi ft)dt \approx \sum_{n=0}^{26} x[n]\sin[2\pi(f/f_s)n]^1, \tag{11.1}$$

where in actuality the basis functions of each of the frequencies are used accordingly. Notice that the given frequencies, 1180 Hz and 980 Hz, are not mutually orthogonal. With a desired symbol rate of 300 Hz, it is necessary for the difference to be half of the symbol rate in order for the signals to be orthogonal. However, as we shall see shortly, in practice, the non-orthongonality does not present a problem. The trignometric values are obtained not from a recursive computation but a table lookup by virtue of

[1] See [12] for a more detailed exposition on the theory of non-coherent demodulation.

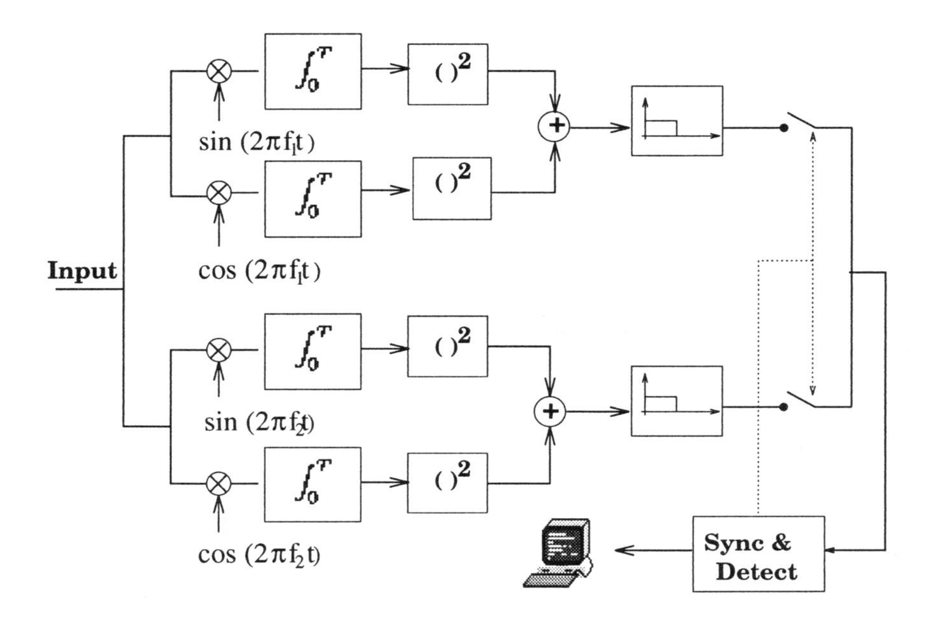

Figure 11.17: Receiver

its availability as mentioned previously. For each frequency, the correlator outputs are squared then summed: $r^2 = r_{sin}^2 + r_{cos}^2$. It is convenient to think of the resulting output as an "envelop," with peaks corresponding to the particular frequency and valleys indicating a different frequency. A graph of the output from a hypothetical input best illustrates this point. See Fig. 11.18. To distinguish between a 0 and a 1 from the two resulting correlator outputs, we take two sequences of 27 samples (300 Hz) from the two correlator outputs, compute the differences between each corresponding samples, and look for the maximum. The position thus found will necessarily be the position at which we wish to interpret the 1s and 0s, since the maximum absolute difference between the two outputs occurs when the "local oscillator" matches as closely in phase as that of the input signal. In effect, the symbol positions will be synchronized.

To reconstruct the transmitted bit pattern, a one or zero, i.e., the output after frequency discrimination, is shifted to the left by 9 and the result added to a running sum of 9-bits. This sum is then shifted to the right by 1. The result is then compared to the 9-bit codeword, 101010101. If the received bits match the codeword, then the next 8 bits will be transferred to the host computer otherwise, the receiver continues to look for the start bits. We mention here that due to the slowness of the direct memory access (DMA), byte transfers to the DSP should not include the start bits.

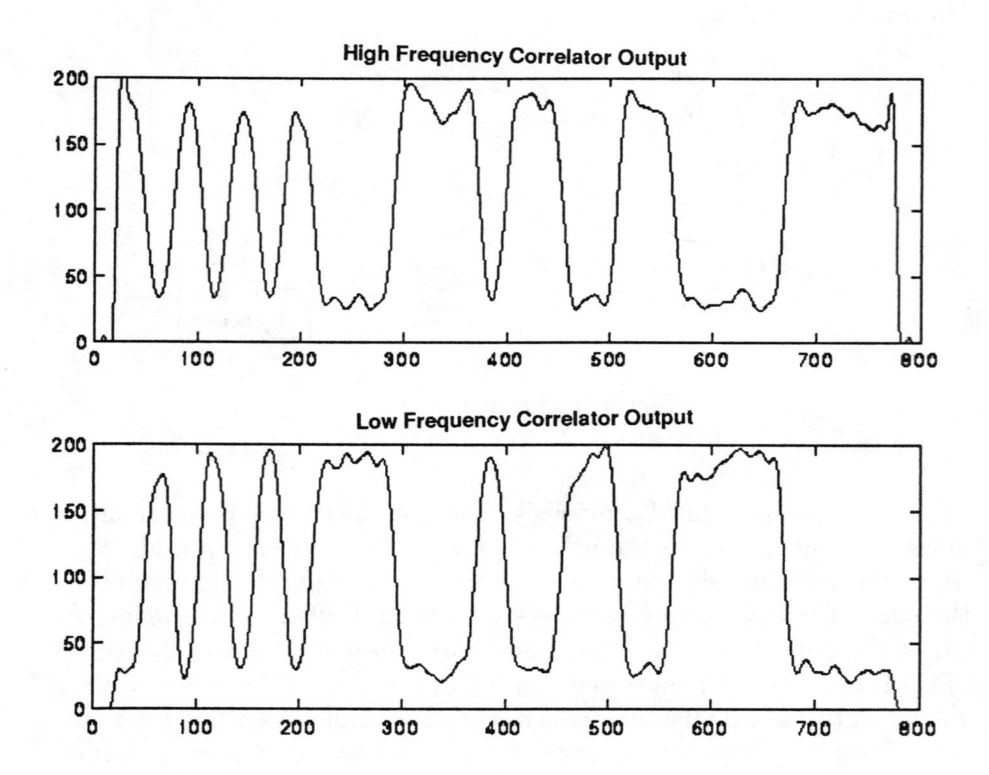

Figure 11.18: Correlator outputs

Rather, the start bits should be hard coded in the modulation program, which places the start bits in the same word as the output byte.

The Programs

The following programs were written on a IBM 286 clone, which is where the DSP resides. The programs can be found on the included disk.

1. `bfsk.asm` — takes input byte from `pc_dsp.c` and modulates it using CPBFSK and attaches a sequence of start bits.
2. `corr.asm` — demodulates signal by non-coherent demodulation. Calls `synch.asm` for symbol synchronization, decodes the received bits, and transfers the resulting byte to `dsp_pc.c`.
3. `fir.asm` — smoothes out the correlator output for decision making.
4. `trigfun.asm` — contains trignometric lookup tables as well as declarations of other buffers.
5. `dma_rcv.asm` — handles byte receives from host computer.
6. `dma_trm.asm` — handles byte transfers to host computer.
7. `pc_dsp.c` — interface program that runs on the host computer, takes input from the keyboard and transfers it to the DSP.
8. `dsp_pc.c` — interface program that runs on the host computer and continuously checks for byte presence on hostport.
9. `pc.c` — supporting subroutines for host interface programs.
10. `pc_1.h` — header file containing various system specific declarations.

Appendix A

Host Data Communications

The communication between the TMS320C30 and the host PC is register-based and consists of a 16-bit communication register and the Test Bus Controller (TBC) on the EVM board [14, 15]. The 16-bit bi-directional register interfaces the lower 8 data bits on the I/O bus of the host PC to the lower 16-bits of the TMS320C30 expansion bus. Table A.1 shows the EVM host interface memory map.

The COM_DATA and COM_CMD locations are the data and command modes of the 16-bit register interface, which are mapped to a single physical register on the EVM board (the C30 host port in Fig. 2.6). The TMS320C30 and the PC exchange data or commands by reading and writing to this register. To distinguish between data (COM_DATA) and commands (COM_CMD), different interrupt levels are used.

The interface is designed such that the TMS320C30 side of the communications is interrupt-driven, while the host side is polled. This means that host accesses to the communications register will generate interrupts to the TMS320C30 and TMS320C30 accesses to the communication register will generate a TBC event flag that the host can poll.

Three interrupt signals are provided to the TMS320C30: INT0, INT1, and INT2. Table A.2 lists the host interrupt signals and describes their purpose. Note that no interrupt results from a host command read. This allows polling of the command word without interrupting the TMS320C30.

Two TBC event flags, EVT1 and EVT2, are used by the host to provide a synchronization mechanism. Table A.3 lists the event signals and describes their function.

Name	PC I/O Base Offset	TMS320C30 Mapping	Host Access
Reserved	0x000 - 0x800	None	Reserved
CONTROL5	0x00A	None	16-bit write
Reserved	0x00C - 0x012	None	Reserved
MINOR_CMD	0x014	None	16-bit read/write
Reserved	0x016 - 0x01E	None	Reserved
STATUS0	0x400	None	16-bit read/write
Reserved	0x402 - 0x41F	None	Reserved
COM_CMD	0x800	0x804000 - 0x805FFF	8-bit read/write
COM_DATA	0x808	0x804000 - 0x805FFF	16-bit read/write
SOFT_RESET	0x818	None	Control point write

Table A.1: TMS320C30 EVM Host Interface Memory Map

Interrupt Pins	Purpose
INT0	Host 8-bit command write to COM_CMD
INT1	Host 16-bit data write to COM_DATA
INT2	Host 16-bit data read to COM_DATA

Table A.2: Host Interrupts to the TMS320C30

Signal	Function
EVT1	Host read acknowledge. This flag is set when the TMS320C30 writes to the communications register.
EVT2	Host write acknowledge. This flag is set when the TMS320C30 reads to the communication register.

Table A.3: EVM Events for the Host

Access	Purpose
Read STATUS0 bit 1	Poll host read acknowledge EVT1
Read STATUS0 bit 2	Poll host write acknowledge EVT2
Write 0x0002 to MINOR_CMD	Clear STATUS0 bit 1
Write 0x0004 to MINOR_CMD	Clear STATUS0 bit 2
Write 0x6044 to MINOR_CMD	Update for next read of STATUS0

Table A.4: Host Verification Register Access

The PC uses the STATUS0 and MINOR_CMD locations for the polling function (see Table A.4). Bit 1 of STATUS0 indicates that the TMS320C30 has performed a write to the communications register and bit 2 indicates that the TMS320C30 has performed a read. The status of these bits is cleared by writing 0x2 and 0x4 to the MINOR_CMD register. However, before reading STATUS0, the status must be updated each time by writing 0x6044 to the MINOR_CMD register.

Thus, by means of a single register, synchronized data transfers between the TMS320C30 and the host can be carried out. The data transfer may also occur as a direct memory access (DMA). The TMS320C30 Evaluation Module Technical Reference [15] contains code for transferring data between the EVM TMS320C30 and the PC host. In the following, we give a simple example, in which we will use a polling scheme on both the TMS320C30 and the host side. This will be instructive for understanding the interface and can be useful for some applications. The example transfers a block of 512 16-bit data elements from the TMS320C30 to the host. Fig. A.1 and A.2 show the core parts of the programs for the TMS320C30 and the PC, respectively.

When the PC is ready to receive data, it sends a start command to the DSP by writing any 8-bit word to the COM_CMD location. This will generate an INT0 interrupt to the TMS320C30, which sets the INT0 flag bit in the IF (Interrupt Flag) register. When the TMS320C30 is ready to send data, it polls the INT0 flag to see if the host is ready to receive. Once the INT0 flag is set, the TMS320C30 starts to send the first data element by writing it into the communications register. This sets the TBC EVT1 flag for the host.

The host, after giving the start command, waits for the TMS320C30 to start sending data. It polls the EVT1 flag by reading and updating the STATUS0 word. Once the EVT1 flag is set, the PC reads the data element from the COM_DATA location and clears the EVT1 flag as an

```
          ldi      @buf_addr, ar0      ;ar0 points to data buffer
          ldi      @hostport, ar1      ;ar1 points to hostport 0x804000
          ldi      511, rc             ;transfer 512 data elements

wait_int0: tstb    1h, if              ;poll host ready int0 flag
          bz       wait_int0           ;wait for int0 flag to be set
          andn     1h, if              ;clear int0 flag

          rptb     loop
          ldi      *ar0++, r0          ;load data
          sti      r0, *ar1            ;send data
wait_int2: tstb    4h, if              ;poll host read int2 flag
          bz       wait_int2           ;wait for int2 flag to be set
loop:     andn     4h, if              ;clear int2 flag
```

Figure A.1: C30 Code for Sending Data to Host PC

```
oupt(COM_CMD, 1);                         /* send 8-bit start command */

for (i = 0; i < 512; i++)                 /* receive 512 data         */
{
   do
   {
      outport(MINOR_CMD, 0x6044);         /* update STATUS0           */
   }
   while(! inport(STATUS0) & 0x0002);     /* poll EVT1 flag           */

   outport(MINOR_CMD, 0x0002);            /* clear EVT1 flag          */
   data_buffer[i] = inport(COM_DATA);     /* read 16-bit data         */
}
```

Figure A.2: PC Code for Receiving Data from TMS320C30 (Turbo C)

acknowledgment. The host read of the COM_DATA register generates an INT2 interrupt to the TMS320C30, which sets the INT2 flag bit in the IF register.

On the other side, the TMS320C30 is polling the INT2 flag waiting for the data element to be read by the host. Once the flag is set, it clears the INT2 flag and writes the second data element to the communication register. Meantime, after reading the first data element and putting it in a buffer, the host goes back to poll the EVT1 waiting to receive the second data element. The process continues in this way until the desired number of data elements has been transferred.

A complete program for a real-time data transfer, based on the above example, is included on the disk in the back of the book. The program is actually a PC oscilloscope, which transfers a block of the most recently received samples from the EVM board to the host PC and plots them on the PC screen.

Appendix B

Interface to C Language

As a high-level language, C is generally faster and easier to program than an assembly language and is portable to a variety of processors, including the TMS320C30 floating-point digital signal processor. The drawback, however, is that the resulting code is less efficient. To reduce the execution time and memory requirements, portions of the code can be developed in an assembly language that are called by a C program. The C compiler for the TMS320C30 is available from Texas Instruments [16]. The compiler imposes a set of conventions for function calls and register uses. These conventions must be followed when you write assembly language functions to interface with C programs. This appendix describes these conventions and gives some examples of assembly language code.

B.1 Register and Calling Conventions

Register Variables

The C compiler uses the following eight registers to store register variables:

R4, R5	used for integer register variables
R6, R7	used for floating-point register variables
AR4 - AR7	used for pointer register variables

These registers are preserved across function calls. An assembly function that uses any of these registers must save the contents of each register used on entrance and be able to restore the contents on exit.

Return Values

The C compiler uses register `R0` to place the return value of a function (integer, pointer, or floating-point). When an assembly function returns a value, the function should put the return value in register `R0`.

Stack and Frame Pointers

The C compiler uses the TMS320C30's built-in stack mechanism to allocate local variables, pass arguments to functions, and save temporary results. Two registers are used to manage the stack:

SP	is the stack pointer; it points to the top of the stack.
AR3	is the frame pointer (`FP`); it points to the beginning of the current local frame.

A local frame is an area on the stack used for storing function arguments and local variables. Each time a function is invoked, a new local frame is created on the top of the stack. The stack pointer (`SP`) points to the top of the stack, i.e., the top of the allocated area. The C compiler uses `AR3` as the frame pointer (`FP`) to the beginning or bottom of the local frame for the current function. All objects stored in the local frame are referenced indirectly through the `FP` register. Fig. B.1 illustrates the use of the stack.

Both the `FP` and the `SP` registers must be preserved across function calls. The `SP` is preserved automatically since everything pushed for a call is popped on return. The `FP` must be preserved specifically at the entry and the exit of a function.

Data Page Pointer

In the C compiler, global and static variables must be stored in an uninitialized section called `.bss`. The compiler supports two memory models that affect how the `.bss` section is allocated in memory: Small Memory Model and Big Memory Model.

The small memory model uses a single 64k memory page for the `.bss` section. This requires that all static and global data and long immediate constants in a program are less than 64k. The compiler sets the Data Page Pointer (`DP`) register at program startup to point to the beginning of the `.bss`. It can then access all the objects in the `.bss` section (global variables, static variables, and constant tables) with direct addressing (`@`) without modifying the `DP`. An assembly language function that changes the `DP` must restore its value.

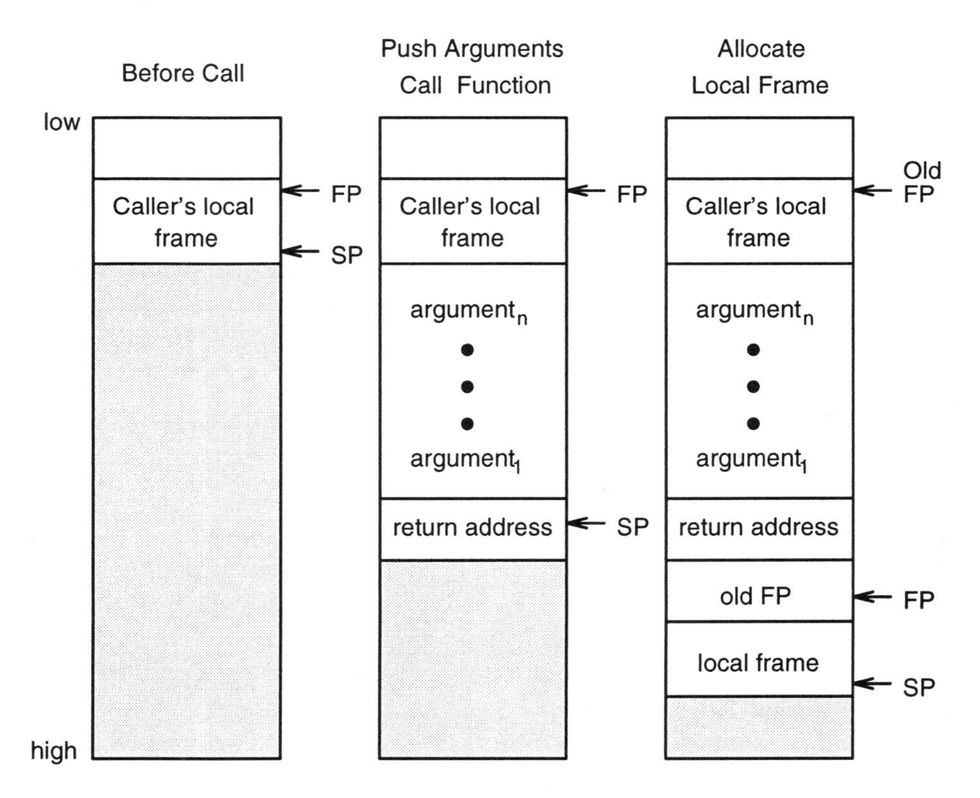

Figure B.1: Stack use during a function call

If the size of `.bss` exceeds 64k, the compiler uses the big memory model. In this model, unlimited space can be used for global and static data, but the compiler must explicitly set the `DP` register each time a global or a static object is accessed. This takes extra instruction words and machine cycles. The `DP` is not preserved for the big memory model.

Global Variables

The C compiler appends an underscore (_) to the beginning of all identifiers. Thus, you must use a prefix of _ in your assembly code modules for all objects that are to be accessible from C.

Any object or function in assembly language that is to be accessed or called from C must be declared with the `.global` directive. This defines the symbol as external and allows the linker to resolve references to it. Likewise, to access a C function or object from assembly, declare the C object with `.global`.

It is sometimes useful for a C program to access variables that are defined in assembly. In this case, besides declaring the variable as `.global` and preceding the name with an underscore, you need to use the `.bss` directive to define the variable in assembly and declare the variable as `extern` in C.

Access of Function Arguments

As discussed before, function arguments and local variables are stored on the stack using a local frame and managed with the stack pointer `SP` and frame pointer `FP` (see Fig. B.1). An assembly function called by a C program accesses its arguments indirectly through the `FP`. Before modifying the `FP`, however, the called function must first save the old `FP` on the stack and set the new `FP` to the current `SP`. The arguments are then addressed with negative offsets from the `FP`. Since the return address is stored at the location directly below the `FP`, the first argument is addressed as `*-FP(2)`. The other arguments are addressed with increasing offsets.

B.2 Example of Assembly Language Function

Fig. B.2 presents an example of calling an assembly language function from C. The assembly function `asmfunc` takes a single argument from C `main` function, adds it to the C global variable `gvar`, and returns the result.

In the assembly language code in Fig. B.2, note the underscores on all the C symbol names. Note that the preserved register `R4` is saved and the return value is placed in register `R0`. Note also that the `DP` needs to be

```
/****************************************************/
/* C Program for interfacing with assembly language */
/****************************************************/

extern int asmfunc();  /* declare external function */
int gvar;              /* declare global variable   */

main()
{
   int i;
   i=asmfunc(i);       /* call asm function normally */
}

;****************************************************
; Assembly Language Program
;****************************************************

FP  .set     AR3           ; use AR3 as FP
    .global  _asmfunc      ; declare external function
    .global  _gvar         ; declare external variable

_asmfunc:
    push     FP            ; save old FP
    push     R4            ; save preserved register R4
    ldi      SP, FP        ; point to top of stack
    ldi      *-FP(2), R4   ; load argument into R4
    ldp      _gvar, DP     ; set DP to page of gvar
                           ; (Big Model Only)
    addi     @_gvar, R4    ; add gvar to argument in R4
    ldi      R4, R0        ; put return value in R0
    pop      R4            ; restore R4
    pop      FP            ; restore FP
    rets
```

Figure B.2: An assembly language function called from C

set only when accessing global variables in the big memory model. For the small memory model, the LDP instruction that loads the page pointer can be omitted.

B.3 Inline Assembly Language

Besides linking separate assembly language modules with C functions, the TMS320C30 C compiler allows you to embed assembly language instructions directly in the C source program. This is realized by using the asm statement. The asm statement is syntactically like a call to a function named asm, with one string-constant argument:

asm("*assembler text* ");

The compiler copies the argument string directly into its output file. The assembler text must be enclosed in double quotes. The inserted code must be a legal assembly language statement. It may be an instruction, directive, or a comment. The asm statement is provided so that you can access the features of the hardware of the TMS320C30, which would be otherwise inaccessible from C. These features include access to the interrupt enable and interrupt flag registers, as well as the status register interrupt enable/disable bit. For example, the following three asm statements function to enable serial port 0 receive interrupt:

```
asm(" ldi     0, IF");  /* clear any interrupt flags               */
asm(" ldi   20h, IE");  /* enable serial port 0 receive interrupt */
asm(" or  2000h, ST");  /* set global interrupt enable bit        */
```

When you use the asm statement, be careful not to disrupt the C environment. The compiler does not check or analyze the inserted assembler instructions.

Appendix C

Matlab Functions

This Appendix lists several Matlab [7] functions and scripts that can be used to check and further explore the experiments in this book. All the functions and scripts can be found on the included disk and require access to Matlab and the Signal Processing Toolbox [7].

C.1 BILINEAR3

```
function [NUMD, DEND] = bilinear3(NUM, DEN, FS, FC, FP)
%BILINEAR3 Bilinear transformation with optional frequency prewarping
%   [NUMD,DEND] = BILINEAR3(NUM,DEN,FS) converts the s-domain trans-
%   fer function specified by NUM and DEN to a z-transform discrete
%   equivalent obtained from the bilinear transformation:
%
%      H(z) = H(s) |
%                  | s = 2*FS*(z-1)/(z+1)
%
%   where NUM and DEN are row vectors containing numerator and deno-
%   minator transfer function coefficients and Fs are the sample fre-
%   quency in Hz.  NUM and DEN are in descending powers of s. Two
%   optional additional input arguments can be used to speficy pre-
%   warping. For example [NUMD,DEND] = BILINEAR3(NUM,DEN,FS,FC,FP)
%   applies prewarping before the bilinear transformation so that
%   the prototype frequency response at FP is mapped to the filter
%   cut off FC. FC and FP are specified in Hz, and the default for
%   FP is 2*pi Hz (=1 rad/s) which is what buttap, cheb1ap, chap2ap,
%   and ellipap generate.
%   Example
%      [z,p,k]=buttap(3); [n,d]=zp2tf(z,p,k);
```

```
%       [n1,d1]=bilinear3(n,d,1000,200);
%       [h,w]=freqz(n1,d1,256); plot(w*500/pi,abs(h))
%    See also BILINEAR
%    Ref: Parks and Burrus: Digital Filter Design, John Wiley, 1987

%Henrik Sorensen, 03-09-1992
%Copyright. 1992-96.

[mn,nn] = size(NUM);
[md,nd] = size(DEN);

if nn > nd
    error('Numerator cannot be higher order than denominator.');
end
if nargin == 4                          % Prewarp
    w = 1/tan(2*pi*FC/FS/2);
elseif nargin == 5                      % Prewarp
    w = 2*pi*FP/tan(2*pi*FC/FS/2);
elseif nargin == 3
    w = 2*FS;
else
    error('Incorrect number of arguments');
end
[a,b,c,d] = tf2ss(NUM,DEN);            % Convert to state-space form
n = max(size(a));                      % Do bilinear x-form
t1 = w*eye(n) + a;
t2 = w*eye(n) - a;
ad = t1/t2;
bd = 2*w*(t2\b);
cd = c/t2;
dd = c/t2*b + d;
DEND = poly(ad);                       % Convert back to transfer function
NUMD = poly(ad-bd*cd)+(dd-1)*DEND;
```

C.2 BUTTCAS

```
function [NUM,DEN,GAIN] = buttcas(N,Fs,Fp)
%BUTTCAS generates length N digital Butterworth filter in cascade form.
%    [NUM,DEN,GAIN]=BUTCAS(N,FS,FP) computes the length N Butterworth
%    digital filter in cascade form using the bilinear transformation
%    with the sampling frequency FS and cut off frequency FP.
%
%    For example to design a third order Butterworth filter with
%    sampling frequency 8kHz and cut off 1kHz:
```

```
%       [n,d,g]=buttcas(3,8000,1000)
%   which gives
%       n=[1 2 1; 0 1 1]
%       d=[1 -1.0448 0.4776;0 1 -0.4142]
%       g=[0.1082 0.2929]
%
%   and hence
%                   z^2 + 2z + 1                     z + 1
%   H(z) = 0.1082*----------------------  *  0.2929*----------
%                   z^2 - 1.0448z + 0.4776           z - 0.4142
%

%   NOTICE: The routine assumes that buttap returns the poles in
%   conjugate pairs in location p(k) and p(N-k).

%Henrik Sorensen, Mar. 6, 1992
%Copyright. 1992-96.

[z,p,g] = buttap(N);         % Get s-plane prototype
                             % Notice z=[] and g=1 always!!
np = length(p);              % Number of poles
np2 = fix(np/2);             % Number of 2nd order sections
fb = np-2*np2;               % First order section

NUM = zeros(np2+fb,3);       % Initialize NUM, DEN, and GAIN
DEN = NUM;
GAIN = zeros(np2+fb,1);

for k=1:np2                  % Compute 2nd order blocks
   [tmp1,DEN(k,:)] = bilinear3([1],[1 -2*real(p(k)) ...
   real(p(k))^2+imag(p(k))^2],Fs,Fp);
   GAIN(k)=tmp1(1);
   NUM(k,:)=tmp1/GAIN(k);
end
if fb==1
   [tmp1,tmp2] = bilinear3([1],[1 -real(p(np2+1))],Fs,Fp);
   GAIN(np2+1) = tmp1(1);
   NUM(np2+1,:) = [0 tmp1/tmp1(1)];
   DEN(np2+1,:) = [0 tmp2];
end
```

C.3 CASBUTTER

```
%CASBUTTER generates the "Cascade Form" coefficients of a length  17
```

```
%   Butterworth filter. Cut-off frequency: 1.2 kHz. Sampling
%   frequency 8 kHz.

%Henrik Sorensen     01/11/92
%Copyright. 1992-96.

%Find numerator, denominator, and gain
[num,den,gain]=buttcas(17,8000,1000);

%Plot filter
n=[1];for i=1:9 n=conv(n,num(i,:)); end;
d=[1];for i=1:9 d=conv(d,den(i,:)); end;
[h,w]=freqz(n,d,200);
plot(w*4000/pi,abs(h));
xlabel('Hz');ylabel('Magnitude');title('Filter table 6.2');

%Print numerator in column form
fprintf('Numerator for second order blocks\n');
fprintf('     b0k                        b1k...
                b2k\n');
%Apply gain to b coefficients
fprintf('%18.12e      %18.12e      %18.12e\n',(num(1:8,:)...
.*kron(gain(1:8),ones(1,3)))');
%Print last row separately to line up properly
fprintf('%18.12e      %18.12e\n\n',num(9,2)*gain(9),num(9,3)*gain(9));

%Print denominator in column form
fprintf('Denominator for second order blocks\n');
fprintf('     a0k                        a1k...
                a2k\n');
%Change sign on a1k and a2k
fprintf('%18.12e      %18.12e      %18.12e\n',(den(1:8,:)...
.*kron(ones(8,1),[1 -1 -1]))');
%Print last row separately to line up properly
fprintf('%18.12e      %18.12e\n\n',den(9,2),-1*den(9,3));
```

C.4 CASCHEBY

```
%CASCHEBY generates the "Cascade Form" coefficients of a length 8
%    type 1 Chebycheb filter with ripple 1.0 dB.

%Henrik Sorensen     01/11/92
%Copyright. 1992-96.
```

```
%Find numerator, denominator, and gain
[num,den,gain]=cheb1cas(8,8000,1000,1.0);
%Print numerator in column form
fprintf('Numerator for second order blocks\n');
fprintf('     b0k                           b1k...
                      b2k\n');
%Apply gain to b coefficients
fprintf('%18.12e      %18.12e      %18.12e\n',...
(num.*kron(gain,ones(1,3)))');
fprintf('\n');

%Print denominator in column form
fprintf('Denominator for second order blocks\n');
fprintf('     a0k                           a1k...
                      a2k\n');
%Change sign on a1k and a2k
fprintf('%18.12e      %18.12e      %18.12e\n',...
(den.*kron(ones(4,1),[1 -1 -1]))');
fprintf('\n');
```

C.5 CHEB1CAS

```
function [NUM,DEN,GAIN] = cheb1cas(N,Fs,Fp,Rp)
%CHEB1CAS makes length N type 1 Chebychev filter in cascade form
%   [NUM,DEN,GAIN]=CHEB1CAS(N,FS,FP,RP) computes a type 1 length N
%   Chebychev digital filter in cascade form using the bilinear
%   transformation. The filter will have ripple RP (in dB), samp-
%   ling frequency FS, and cut off frequency FP.
%
%   For example to design a third order Chebychev filter with 0.5 dB
%   ripple, sampling frequency 8kHz, and cut off 1kHz:
%      [n,d,g]=cheb1cas(3,8000,1000,0.5)
%   which gives
%      n=[1 2 1; 0 1 1]
%      d=[1 0.1029 0.5475; 0 1 -0.2297]
%      g=[0.3612 0.6148]
%
%   and hence
%                   z^2 + 2z + 1                     z + 1
%   H(z) = 0.3612*---------------------- *  0.6148*------------
%                   z^2 + 0.1029z + 0.5475           z - 0.2297
%

%   NOTICE: The routine assumes that cheb1ap returns the poles in
```

```
%   conjugate pairs in location p(k) and p(N-k).

%Henrik Sorensen, Mar. 6, 1992
%Copyright. 1992-96.

[z,p,g] = cheb1ap(N,Rp);     % Get s-plane prototype
                             % Notice z=[] and g=1 always!!
np = length(p);              % Number of poles
np2 = fix(np/2);             % Number of 2nd order sections
fb = np-2*np2;               % First order section

NUM = zeros(np2+fb,3);       % Initialize NUM, DEN, and GAIN
DEN = NUM;
GAIN = zeros(np2+fb,1);

for k=1:np2                  % Compute 2nd order blocks
   [tmp1,DEN(k,:)] = bilinear3([1],[1 -2*real(p(k)) ...
   real(p(k))^2+imag(p(k))^2],Fs,Fp);
   GAIN(k)=tmp1(1);
   NUM(k,:)=tmp1/GAIN(k);
end
if fb==1
   [tmp1,tmp2] = bilinear3([1],[1 -real(p(np2+1))],Fs,Fp);
   GAIN(np2+1) = tmp1(1);
   NUM(np2+1,:) = [0 tmp1/tmp1(1)];
   DEN(np2+1,:) = [0 tmp2];
end
```

C.6 DFBUTTER

```
%DFBUTTER generates the "Direct Form" coefficients of a length 17
%   Butterworth filter. Cut-off frequency: 1.2 kHz. Sampling
%   frequency 8 kHz.

%Henrik Sorensen     01/11/92
%Copyright. 1992-96.

%Find numerator and denominator
[num,den]=butter(17,(2*1000/8000));

%Plot filter response
[h,w]=freqz(num,den,200);
plot(w*4000/pi,abs(h));
xlabel('Hz');ylabel('Magnitude');title('Filter table 6.1');
```

```
%Print numerator in column form
fprintf('Numerator\n');fprintf('b(0) = %18.12e\n',num(1));
for i=2:18
   fprintf('b(%d) = b(0) * %g = %18.12e\n',i-1,num(i)/num(1),num(i));
end
fprintf('\n');

%Print denominator in column form
fprintf('Denominator\n');fprintf('a(0) = %18.12e\n',den(1));
for i=2:18
   fprintf('a(%d) = %18.12e\n',i-1,-1*den(i));
end
fprintf('\n');
```

C.7 DITFFT

```
function [YR,YI]=ditfft(XR,XI,M)
%DITFFT computes a radix-2 decimation-in-time FFT.
%   [YR,YI]]=DITFFT(XR,XI,M) computes a length 2^M decimation-
%   in-time radix-2 FFT. The input and output is (XR+jXI).
%
%   For example to compute the length 8 FFT of the sequence
%   [1 5 3 4 2 7 8 6]:
%      [r,i]=ditfft([1 5 3 4 2 7 8 6],[0 0 0 0 0 0 0 0],3)
%   which gives
%      r = [36.00 -1.00 -8.00 -1.00 -8.00 -1.00 -8.00 -1.00]
%      i = [ 0.00  7.83 -2.00 -2.17  0.00  2.17  2.00 -7.83]

%H.V. Sorensen, Jan. 05, 1985
%Copyright. 1985-96.

% Transform length. Notice; can be deduced as max(size(xr))
n=2^M;

% Digit reverse counter. Note this can be done using bitrev command
j = 1; n1 = n - 1;
for i = 1:n1
   if i < j,
      xt    = XR(j); XR(j) = XR(i); XR(i) = xt;    % Swap real part
      xt    = XI(j); XI(j) = XI(i); XI(i) = xt;    % Swap imag part
   end
   k = n/2;
   while k < j,
```

```
      j = j - k; k = k/2;
   end
   j = j + k;
end

% Main FFT loops; computed in-place
n1 = 1;
for k=1:M
   n2 = n1; n1 = n2*2;
   e  = 6.283185307179586/n1;
   a  = 0;
   for j=1:n2
      C = cos(a); S = sin(a);
      a = j*e;
      for i=j:n1:n
         l = i + n2;
         xt    = C*XR(l) + S*XI(l);
         yt    = C*XI(l) - S*XR(l);
         XR(l) = XR(i)  - xt;
         XR(i) = XR(i)  + xt;
         XI(l) = XI(i)  - yt;
         XI(i) = XI(i)  + yt;
      end
   end
end

%Calculation done in-place. Copy result.
YR=XR; YI=XI;
```

C.8 FIRBOX

```
%FIRBOX generates FIR filter using the window design
%   technique with a boxcar window.

%Henrik Sorensen     31/01/93
%Copyright. 1993-96.

%Get filter order
% N=25 in Exp. 4.2.1.1  and N=63 in Exp. 4.2.1.2
N=input('Filter order: ')

%Design filter
a=fir1(N-1,2*1200/8000,boxcar(N))';
firplot(a,200);
```

```
%Compute and print filter coefficients
fprintf('Filter  coefficients:\n');
for i=1:N
   fprintf('h(%d) = %18.12e\n',i-1,a(i));
end
```

C.9 FIRHAM

```
%FIRHAM generates FIR filter using the window design
%   technique with a Hamming window.

%Henrik Sorensen     31/01/93
%Copyright. 1993-96.

%Get filter order
% N=25 in Exp. 4.2.1.1  and N=63 in Exp. 4.2.1.2
N=input('Filter order: ');

%Design filter
a=fir1(N-1,[2*1000/8000 2*2000/8000],hamming(N))';
firplot(a,200);

%Compute and print filter coefficients
fprintf('Filter  coefficients:\n');
for i=1:N
   fprintf('h(%d) = %18.12e\n',i-1,a(i));
end
```

C.10 FIRREM

```
%FIRREM generates FIR filter using the Parks-McClellan
%   (Remez Exchange) design method.

%Henrik Sorensen     31/01/93
%Copyright. 1993-96.

%Filter order
N=42;

%Sampling frequency
Fs=8000;

%Find F and M to call REMEZ
```

```
F=[0.0  0.2  0.25  0.5  0.6  1.0];
M=[0.0  0.0  1.0   1.0  0.0  0.0];
a=remez(N-1,F,M);
firplot(a,200);

%Compute and print filter coefficients
fprintf('Filter  coefficients:\n');
for i=1:N
   fprintf('h(%d) = %18.12e\n',i-1,a(i));
end
```

C.11 FIRPLOT

```
function firplot(H,N,Fs)
%FIRPLOT Plots various responses of a FIR filter
%   FIRPLOT(H,N) Plots impulse, frequency response of a FIR filter
%   with impulse response H. The plots have N points of
%   resolution; if N is not given N=100 is used.
%   The sampling frequency Fs, can be specified such that plots
%   are appropriately labeled.
%
%   For example
%      h=remez(9,[0 0.15 0.2 1],[1 1 0 0]);
%      h=firplot(h,200);

% H. Sorensen, Apr 17, 90, Modified: 4/11/91
%Copyright. 1992-96.

subplot(4,1,1);
stem(H);
title('Impulse response');
xlabel('Filter index');
ylabel('Magnitude')
if nargin == 1
   N = 100;
   fprintf('\nFIRPLOT: N not specified. Frequency plots made ...
   with 100 points');
end
[a,w]=freqz(H,1,N);
if nargin == 3        % Fs specified
   w=Fs*w/pi/2;
end
ma=abs(a);pa=angle(a);
subplot(4,1,2);plot(w,ma);
```

```
title('Magnitude plot');
xlabel('Frequency');
ylabel('Magnitude')
subplot(4,1,3);plot(w,pa);
title('Phase plot');
xlabel('Frequency');
ylabel('Angle (rad)');
subplot(4,1,4);semilogy(w,ma);
title('Logarithmic magnitude plot');
xlabel('Frequency');
ylabel('dB');
```

C.12 NOISEGEN

```
function [Y] = noisegen(N,M,V,X0,K)
%NOISEGEN generates N gaussian distributed random numbers.
%   [Y]=NOISEGEN(N,M,V,X0,K) generates N gaussian distributed
%   random variables with mean M and variance V. X0 is initial
%   seed to random number generator. K is the number of random
%   numbers used to generate each gaussian variable.
%   Example:
%      [Y]=noisegen(100,1,1,pi,12)

%   NOTICE: Not written for optimal Matlab performance,
%   but to illustrate C30 implementation.

%Henrik Sorensen, Mar. 6, 1992
%Copyright. 1992-96.

for i=1:N
   [X,X1]=randgen(K,X0);      % Get K random numbers
   X0=X1;
   Y(i)=M+V*(sum(X)*4.6566128752457e-10-K/2)/sqrt(K/12);
end
```

C.13 R2FFT

```
function [YR,YI] = r2fft(XR,M)
%R2FFT computes FFT of real-data input sequence
%   [YR,YI]=R2FFT(XR,M) Computes the length 2^M DFT of the real
%   sequence XR using a Cooley-Tukey radix-2 DIT in-place FFT.
%   Output contains symmetries.
%
%   For example to compute the length 8 FFT of the sequence
```

```
%   [1 5 3 4 2 7 8 6]:
%      [r,i]=r2fft([1 5 3 4 2 7 8 6],3)
%   which gives
%      r = [36.00 -1.00 -8.00 -1.00 -8.00 -1.00 -8.00 -1.00]
%      i = [ 0.00  7.83 -2.00 -2.17  0.00  2.17  2.00 -7.83]

%H.V. Sorensen,    Jan. 1985
%Copyright. 1992-96.

%Transform length. Notice; can be deduced as: n=max(size(XR))
n = 2^M;

% Digit reverse counter. Note this can be done using bitrev command
j = 1; n1 = n - 1;
for i = 1:n1
   if i < j,
      xt    = XR(j); XR(j) = XR(i); XR(i) = xt;   % Swap
   end
   k = n/2;
   while k < j,
      j = j - k; k = k/2;
   end
   j = j + k;
end


%Length two butterflies
for i=1:2:n
   xt = XR(i);
   XR(i) = xt + XR(i+1);
   XR(i+1) = xt - XR(i+1);
end
n2 = 1;
for k=2:M
   n4 = n2; n2 = 2*n4; n1 = 2*n2;
   e = 6.283185307179586/n1;
   for i=1:n1:n
      xt = XR(i);
      XR(i) = xt + XR(i+n2);
      XR(i+n2) = xt - XR(i+n2);
      XR(i+n4+n2) = -XR(i+n4+n2);
      a = e;
      for j=1:n4-1
         i1=i+j;i2=i-j+n2;i3=i+j+n2;i4=i-j+n1;
         cc = cos(a);ss = sin(a);
```

```
            a = a + e;
            t1 = XR(i3)*cc+XR(i4)*ss;
            t2 = XR(i3)*ss-XR(i4)*cc;
            XR(i4) =  XR(i2) - t2;
            XR(i3) = -XR(i2) - t2;
            XR(i2) =  XR(i1) - t1;
            XR(i1) =  XR(i1) + t1;
         end
      end
   end

%Calculation done in-place. Extract result.
n1=2^(M-1);
a=size(XR);
if a(1)<a(2)
   YR=[XR(1:n1+1) XR(n1:-1:2)];
   YI=[0 XR(n:-1:n1+2) 0 -XR(n1+2:n)];
else
   YR=[XR(1:n1+1);XR(n1:-1:2)];
   YI=[0;XR(n:-1:n1+2);0;-XR(n1+2:n)];
end
return
```

C.14 RANDGEN

```
function [Y,X1] = randgen(N,X0)
%RANDGEN generates N random numbers.
%   [Y,X1]=RANDGEN(N,X0) generates N random numbers with initial
%   seed X0. X1 is the final seed of generator.
%   Example:
%       [Y]=randgen(100,pi)

%   NOTICE: Not written for optimal Matlab performance,
%   but to illustrate C30 implementation.

%Henrik Sorensen, Mar. 6, 1992
%Copyright. 1992-96.

Y=X0;                               % x[n]=X0
for i=2:N
% mod() operation needed since Matlab does not perform 32-bit
% multiplecations like the C30
   xpn1=rem(Y(i-1)*65539,2^32);     % x'[n+1]=x[n]*65539
   if xpn1<2^31                     % if x'[n+1]>=0
```

```
      Y(i)=xpn1;                        %      x[n+1]=x'[n+1]
   else                                 % else
      Y(i)=xpn1-2^31;                   %      x[n+1]=x'[n+1]-2^31
   end
end
X1=Y(N);
```

C.15 SINGEN

```
function [Y] = singen(N,F,Fs)
%SINGEN generates N samples of sinewave.
%   [Y]=SINGEN(N,F,FS) generates N samples of sinewave with
%   frequency F at sampling frequency Fs.
%   Example:
%      [Y]=singen(100,1000,8000);
%      plot(Y);

%   NOTICE: Not written for optimal Matlab performance,
%   but to illustrate C30 implementation.

%Henrik Sorensen, Mar. 6, 1992
%Copyright. 1992-96.

A1=2*cos(2*pi*F/Fs); % Constant which controls frequency of sinewave
y1=1;                % y[1]=1 (from delta function)
y2=0;                % y[0]=0
for i=1:N
   yn=A1*y1-y2;      % y[n]= delta[n]+A1*y[n-1]+A0*y[n-2]
   y2=y1;            % y[n-2]=y[n-1]
   y1=yn;            % y[n-1]=y[n]
   Y(i)=yn;          % Put y[n] to output
end
```

Appendix D

Miscellaneous Programs

This Appendix lists several TMS320C30 routines and PC programs that are used in the course, but does not "fit" into the regular text.

D.1 TWF512.ASM

```
******************************************************************************
*                                                                            *
*                      Twiddle factors for FFT - TWF512.ASM                  *
*                      ------------------------------------                  *
*                                                                            *
*   Table of twiddle factors for 512 point FFT                               *
*                                                                            *
*      Henrik Sorensen.                                                      *
*      Written in Aug. 12, 1992. Modified in Feb 10, 1995. Version 1.0       *
*                                                                            *
*  Do not distribute without permission from authors. Copyright 1995-96      *
*                                                                            *
******************************************************************************
           .global SINE
           .global N
           .global M

           .data
SINE
           .float  0.00000000e+00
           .float  1.22715383e-02
           .float  2.45412285e-02
           .float  3.68072229e-02
```

```
.float 4.90676743e-02
.float 6.13207363e-02
.float 7.35645636e-02
.float 8.57973123e-02
.float 9.80171403e-02
.float 1.10222207e-01
.float 1.22410675e-01
.float 1.34580709e-01
.float 1.46730474e-01
.float 1.58858143e-01
.float 1.70961889e-01
.float 1.83039888e-01
.float 1.95090322e-01
.float 2.07111376e-01
.float 2.19101240e-01
.float 2.31058108e-01
.float 2.42980180e-01
.float 2.54865660e-01
.float 2.66712757e-01
.float 2.78519689e-01
.float 2.90284677e-01
.float 3.02005949e-01
.float 3.13681740e-01
.float 3.25310292e-01
.float 3.36889853e-01
.float 3.48418680e-01
.float 3.59895037e-01
.float 3.71317194e-01
.float 3.82683432e-01
.float 3.93992040e-01
.float 4.05241314e-01
.float 4.16429560e-01
.float 4.27555093e-01
.float 4.38616239e-01
.float 4.49611330e-01
.float 4.60538711e-01
.float 4.71396737e-01
.float 4.82183772e-01
.float 4.92898192e-01
.float 5.03538384e-01
.float 5.14102744e-01
.float 5.24589683e-01
.float 5.34997620e-01
.float 5.45324988e-01
.float 5.55570233e-01
```

```
        .float  5.65731811e-01
        .float  5.75808191e-01
        .float  5.85797857e-01
        .float  5.95699304e-01
        .float  6.05511041e-01
        .float  6.15231591e-01
        .float  6.24859488e-01
        .float  6.34393284e-01
        .float  6.43831543e-01
        .float  6.53172843e-01
        .float  6.62415778e-01
        .float  6.71558955e-01
        .float  6.80600998e-01
        .float  6.89540545e-01
        .float  6.98376249e-01
        .float  7.07106781e-01
        .float  7.15730825e-01
        .float  7.24247083e-01
        .float  7.32654272e-01
        .float  7.40951125e-01
        .float  7.49136395e-01
        .float  7.57208847e-01
        .float  7.65167266e-01
        .float  7.73010453e-01
        .float  7.80737229e-01
        .float  7.88346428e-01
        .float  7.95836905e-01
        .float  8.03207531e-01
        .float  8.10457198e-01
        .float  8.17584813e-01
        .float  8.24589303e-01
        .float  8.31469612e-01
        .float  8.38224706e-01
        .float  8.44853565e-01
        .float  8.51355193e-01
        .float  8.57728610e-01
        .float  8.63972856e-01
        .float  8.70086991e-01
        .float  8.76070094e-01
        .float  8.81921264e-01
        .float  8.87639620e-01
        .float  8.93224301e-01
        .float  8.98674466e-01
        .float  9.03989293e-01
        .float  9.09167983e-01
```

```
            .float  9.14209756e-01
            .float  9.19113852e-01
            .float  9.23879533e-01
            .float  9.28506080e-01
            .float  9.32992799e-01
            .float  9.37339012e-01
            .float  9.41544065e-01
            .float  9.45607325e-01
            .float  9.49528181e-01
            .float  9.53306040e-01
            .float  9.56940336e-01
            .float  9.60430519e-01
            .float  9.63776066e-01
            .float  9.66976471e-01
            .float  9.70031253e-01
            .float  9.72939952e-01
            .float  9.75702130e-01
            .float  9.78317371e-01
            .float  9.80785280e-01
            .float  9.83105487e-01
            .float  9.85277642e-01
            .float  9.87301418e-01
            .float  9.89176510e-01
            .float  9.90902635e-01
            .float  9.92479535e-01
            .float  9.93906970e-01
            .float  9.95184727e-01
            .float  9.96312612e-01
            .float  9.97290457e-01
            .float  9.98118113e-01
            .float  9.98795456e-01
            .float  9.99322385e-01
            .float  9.99698819e-01
            .float  9.99924702e-01
COS:        .float  1.00000000e+00
            .float  9.99924702e-01
            .float  9.99698819e-01
            .float  9.99322385e-01
            .float  9.98795456e-01
            .float  9.98118113e-01
            .float  9.97290457e-01
            .float  9.96312612e-01
            .float  9.95184727e-01
            .float  9.93906970e-01
            .float  9.92479535e-01
```

```
        .float  9.90902635e-01
        .float  9.89176510e-01
        .float  9.87301418e-01
        .float  9.85277642e-01
        .float  9.83105487e-01
        .float  9.80785280e-01
        .float  9.78317371e-01
        .float  9.75702130e-01
        .float  9.72939952e-01
        .float  9.70031253e-01
        .float  9.66976471e-01
        .float  9.63776066e-01
        .float  9.60430519e-01
        .float  9.56940336e-01
        .float  9.53306040e-01
        .float  9.49528181e-01
        .float  9.45607325e-01
        .float  9.41544065e-01
        .float  9.37339012e-01
        .float  9.32992799e-01
        .float  9.28506080e-01
        .float  9.23879533e-01
        .float  9.19113852e-01
        .float  9.14209756e-01
        .float  9.09167983e-01
        .float  9.03989293e-01
        .float  8.98674466e-01
        .float  8.93224301e-01
        .float  8.87639620e-01
        .float  8.81921264e-01
        .float  8.76070094e-01
        .float  8.70086991e-01
        .float  8.63972856e-01
        .float  8.57728610e-01
        .float  8.51355193e-01
        .float  8.44853565e-01
        .float  8.38224706e-01
        .float  8.31469612e-01
        .float  8.24589303e-01
        .float  8.17584813e-01
        .float  8.10457198e-01
        .float  8.03207531e-01
        .float  7.95836905e-01
        .float  7.88346428e-01
        .float  7.80737229e-01
```

```
.float  7.73010453e-01
.float  7.65167266e-01
.float  7.57208847e-01
.float  7.49136395e-01
.float  7.40951125e-01
.float  7.32654272e-01
.float  7.24247083e-01
.float  7.15730825e-01
.float  7.07106781e-01
.float  6.98376249e-01
.float  6.89540545e-01
.float  6.80600998e-01
.float  6.71558955e-01
.float  6.62415778e-01
.float  6.53172843e-01
.float  6.43831543e-01
.float  6.34393284e-01
.float  6.24859488e-01
.float  6.15231591e-01
.float  6.05511041e-01
.float  5.95699304e-01
.float  5.85797857e-01
.float  5.75808191e-01
.float  5.65731811e-01
.float  5.55570233e-01
.float  5.45324988e-01
.float  5.34997620e-01
.float  5.24589683e-01
.float  5.14102744e-01
.float  5.03538384e-01
.float  4.92898192e-01
.float  4.82183772e-01
.float  4.71396737e-01
.float  4.60538711e-01
.float  4.49611330e-01
.float  4.38616239e-01
.float  4.27555093e-01
.float  4.16429560e-01
.float  4.05241314e-01
.float  3.93992040e-01
.float  3.82683432e-01
.float  3.71317194e-01
.float  3.59895037e-01
.float  3.48418680e-01
.float  3.36889853e-01
```

```
        .float  3.25310292e-01
        .float  3.13681740e-01
        .float  3.02005949e-01
        .float  2.90284677e-01
        .float  2.78519689e-01
        .float  2.66712757e-01
        .float  2.54865660e-01
        .float  2.42980180e-01
        .float  2.31058108e-01
        .float  2.19101240e-01
        .float  2.07111376e-01
        .float  1.95090322e-01
        .float  1.83039888e-01
        .float  1.70961889e-01
        .float  1.58858143e-01
        .float  1.46730474e-01
        .float  1.34580709e-01
        .float  1.22410675e-01
        .float  1.10222207e-01
        .float  9.80171403e-02
        .float  8.57973123e-02
        .float  7.35645636e-02
        .float  6.13207363e-02
        .float  4.90676743e-02
        .float  3.68072229e-02
        .float  2.45412285e-02
        .float  1.22715383e-02
        .float  1.22460635e-16
        .float  -1.22715383e-02
        .float  -2.45412285e-02
        .float  -3.68072229e-02
        .float  -4.90676743e-02
        .float  -6.13207363e-02
        .float  -7.35645636e-02
        .float  -8.57973123e-02
        .float  -9.80171403e-02
        .float  -1.10222207e-01
        .float  -1.22410675e-01
        .float  -1.34580709e-01
        .float  -1.46730474e-01
        .float  -1.58858143e-01
        .float  -1.70961889e-01
        .float  -1.83039888e-01
        .float  -1.95090322e-01
        .float  -2.07111376e-01
```

```
.float  -2.19101240e-01
.float  -2.31058108e-01
.float  -2.42980180e-01
.float  -2.54865660e-01
.float  -2.66712757e-01
.float  -2.78519689e-01
.float  -2.90284677e-01
.float  -3.02005949e-01
.float  -3.13681740e-01
.float  -3.25310292e-01
.float  -3.36889853e-01
.float  -3.48418680e-01
.float  -3.59895037e-01
.float  -3.71317194e-01
.float  -3.82683432e-01
.float  -3.93992040e-01
.float  -4.05241314e-01
.float  -4.16429560e-01
.float  -4.27555093e-01
.float  -4.38616239e-01
.float  -4.49611330e-01
.float  -4.60538711e-01
.float  -4.71396737e-01
.float  -4.82183772e-01
.float  -4.92898192e-01
.float  -5.03538384e-01
.float  -5.14102744e-01
.float  -5.24589683e-01
.float  -5.34997620e-01
.float  -5.45324988e-01
.float  -5.55570233e-01
.float  -5.65731811e-01
.float  -5.75808191e-01
.float  -5.85797857e-01
.float  -5.95699304e-01
.float  -6.05511041e-01
.float  -6.15231591e-01
.float  -6.24859488e-01
.float  -6.34393284e-01
.float  -6.43831543e-01
.float  -6.53172843e-01
.float  -6.62415778e-01
.float  -6.71558955e-01
.float  -6.80600998e-01
.float  -6.89540545e-01
```

```
        .float  -6.98376249e-01
        .float  -7.07106781e-01
        .float  -7.15730825e-01
        .float  -7.24247083e-01
        .float  -7.32654272e-01
        .float  -7.40951125e-01
        .float  -7.49136395e-01
        .float  -7.57208847e-01
        .float  -7.65167266e-01
        .float  -7.73010453e-01
        .float  -7.80737229e-01
        .float  -7.88346428e-01
        .float  -7.95836905e-01
        .float  -8.03207531e-01
        .float  -8.10457198e-01
        .float  -8.17584813e-01
        .float  -8.24589303e-01
        .float  -8.31469612e-01
        .float  -8.38224706e-01
        .float  -8.44853565e-01
        .float  -8.51355193e-01
        .float  -8.57728610e-01
        .float  -8.63972856e-01
        .float  -8.70086991e-01
        .float  -8.76070094e-01
        .float  -8.81921264e-01
        .float  -8.87639620e-01
        .float  -8.93224301e-01
        .float  -8.98674466e-01
        .float  -9.03989293e-01
        .float  -9.09167983e-01
        .float  -9.14209756e-01
        .float  -9.19113852e-01
        .float  -9.23879533e-01
        .float  -9.28506080e-01
        .float  -9.32992799e-01
        .float  -9.37339012e-01
        .float  -9.41544065e-01
        .float  -9.45607325e-01
        .float  -9.49528181e-01
        .float  -9.53306040e-01
        .float  -9.56940336e-01
        .float  -9.60430519e-01
        .float  -9.63776066e-01
        .float  -9.66976471e-01
```

```
.float  -9.70031253e-01
.float  -9.72939952e-01
.float  -9.75702130e-01
.float  -9.78317371e-01
.float  -9.80785280e-01
.float  -9.83105487e-01
.float  -9.85277642e-01
.float  -9.87301418e-01
.float  -9.89176510e-01
.float  -9.90902635e-01
.float  -9.92479535e-01
.float  -9.93906970e-01
.float  -9.95184727e-01
.float  -9.96312612e-01
.float  -9.97290457e-01
.float  -9.98118113e-01
.float  -9.98795456e-01
.float  -9.99322385e-01
.float  -9.99698819e-01
.float  -9.99924702e-01
.float  -1.00000000e+00
.float  -9.99924702e-01
.float  -9.99698819e-01
.float  -9.99322385e-01
.float  -9.98795456e-01
.float  -9.98118113e-01
.float  -9.97290457e-01
.float  -9.96312612e-01
.float  -9.95184727e-01
.float  -9.93906970e-01
.float  -9.92479535e-01
.float  -9.90902635e-01
.float  -9.89176510e-01
.float  -9.87301418e-01
.float  -9.85277642e-01
.float  -9.83105487e-01
.float  -9.80785280e-01
.float  -9.78317371e-01
.float  -9.75702130e-01
.float  -9.72939952e-01
.float  -9.70031253e-01
.float  -9.66976471e-01
.float  -9.63776066e-01
.float  -9.60430519e-01
.float  -9.56940336e-01
```

```
.float  -9.53306040e-01
.float  -9.49528181e-01
.float  -9.45607325e-01
.float  -9.41544065e-01
.float  -9.37339012e-01
.float  -9.32992799e-01
.float  -9.28506080e-01
.float  -9.23879533e-01
.float  -9.19113852e-01
.float  -9.14209756e-01
.float  -9.09167983e-01
.float  -9.03989293e-01
.float  -8.98674466e-01
.float  -8.93224301e-01
.float  -8.87639620e-01
.float  -8.81921264e-01
.float  -8.76070094e-01
.float  -8.70086991e-01
.float  -8.63972856e-01
.float  -8.57728610e-01
.float  -8.51355193e-01
.float  -8.44853565e-01
.float  -8.38224706e-01
.float  -8.31469612e-01
.float  -8.24589303e-01
.float  -8.17584813e-01
.float  -8.10457198e-01
.float  -8.03207531e-01
.float  -7.95836905e-01
.float  -7.88346428e-01
.float  -7.80737229e-01
.float  -7.73010453e-01
.float  -7.65167266e-01
.float  -7.57208847e-01
.float  -7.49136395e-01
.float  -7.40951125e-01
.float  -7.32654272e-01
.float  -7.24247083e-01
.float  -7.15730825e-01
.float  -7.07106781e-01
.float  -6.98376249e-01
.float  -6.89540545e-01
.float  -6.80600998e-01
.float  -6.71558955e-01
.float  -6.62415778e-01
```

```
.float  -6.53172843e-01
.float  -6.43831543e-01
.float  -6.34393284e-01
.float  -6.24859488e-01
.float  -6.15231591e-01
.float  -6.05511041e-01
.float  -5.95699304e-01
.float  -5.85797857e-01
.float  -5.75808191e-01
.float  -5.65731811e-01
.float  -5.55570233e-01
.float  -5.45324988e-01
.float  -5.34997620e-01
.float  -5.24589683e-01
.float  -5.14102744e-01
.float  -5.03538384e-01
.float  -4.92898192e-01
.float  -4.82183772e-01
.float  -4.71396737e-01
.float  -4.60538711e-01
.float  -4.49611330e-01
.float  -4.38616239e-01
.float  -4.27555093e-01
.float  -4.16429560e-01
.float  -4.05241314e-01
.float  -3.93992040e-01
.float  -3.82683432e-01
.float  -3.71317194e-01
.float  -3.59895037e-01
.float  -3.48418680e-01
.float  -3.36889853e-01
.float  -3.25310292e-01
.float  -3.13681740e-01
.float  -3.02005949e-01
.float  -2.90284677e-01
.float  -2.78519689e-01
.float  -2.66712757e-01
.float  -2.54865660e-01
.float  -2.42980180e-01
.float  -2.31058108e-01
.float  -2.19101240e-01
.float  -2.07111376e-01
.float  -1.95090322e-01
.float  -1.83039888e-01
.float  -1.70961889e-01
```

```
.float  -1.58858143e-01
.float  -1.46730474e-01
.float  -1.34580709e-01
.float  -1.22410675e-01
.float  -1.10222207e-01
.float  -9.80171403e-02
.float  -8.57973123e-02
.float  -7.35645636e-02
.float  -6.13207363e-02
.float  -4.90676743e-02
.float  -3.68072229e-02
.float  -2.45412285e-02
.float  -1.22715383e-02
.float  -2.44921271e-16
.float  1.22715383e-02
.float  2.45412285e-02
.float  3.68072229e-02
.float  4.90676743e-02
.float  6.13207363e-02
.float  7.35645636e-02
.float  8.57973123e-02
.float  9.80171403e-02
.float  1.10222207e-01
.float  1.22410675e-01
.float  1.34580709e-01
.float  1.46730474e-01
.float  1.58858143e-01
.float  1.70961889e-01
.float  1.83039888e-01
.float  1.95090322e-01
.float  2.07111376e-01
.float  2.19101240e-01
.float  2.31058108e-01
.float  2.42980180e-01
.float  2.54865660e-01
.float  2.66712757e-01
.float  2.78519689e-01
.float  2.90284677e-01
.float  3.02005949e-01
.float  3.13681740e-01
.float  3.25310292e-01
.float  3.36889853e-01
.float  3.48418680e-01
.float  3.59895037e-01
.float  3.71317194e-01
```

```
.float 3.82683432e-01
.float 3.93992040e-01
.float 4.05241314e-01
.float 4.16429560e-01
.float 4.27555093e-01
.float 4.38616239e-01
.float 4.49611330e-01
.float 4.60538711e-01
.float 4.71396737e-01
.float 4.82183772e-01
.float 4.92898192e-01
.float 5.03538384e-01
.float 5.14102744e-01
.float 5.24589683e-01
.float 5.34997620e-01
.float 5.45324988e-01
.float 5.55570233e-01
.float 5.65731811e-01
.float 5.75808191e-01
.float 5.85797857e-01
.float 5.95699304e-01
.float 6.05511041e-01
.float 6.15231591e-01
.float 6.24859488e-01
.float 6.34393284e-01
.float 6.43831543e-01
.float 6.53172843e-01
.float 6.62415778e-01
.float 6.71558955e-01
.float 6.80600998e-01
.float 6.89540545e-01
.float 6.98376249e-01
.float 7.07106781e-01
.float 7.15730825e-01
.float 7.24247083e-01
.float 7.32654272e-01
.float 7.40951125e-01
.float 7.49136395e-01
.float 7.57208847e-01
.float 7.65167266e-01
.float 7.73010453e-01
.float 7.80737229e-01
.float 7.88346428e-01
.float 7.95836905e-01
.float 8.03207531e-01
```

```
.float  8.10457198e-01
.float  8.17584813e-01
.float  8.24589303e-01
.float  8.31469612e-01
.float  8.38224706e-01
.float  8.44853565e-01
.float  8.51355193e-01
.float  8.57728610e-01
.float  8.63972856e-01
.float  8.70086991e-01
.float  8.76070094e-01
.float  8.81921264e-01
.float  8.87639620e-01
.float  8.93224301e-01
.float  8.98674466e-01
.float  9.03989293e-01
.float  9.09167983e-01
.float  9.14209756e-01
.float  9.19113852e-01
.float  9.23879533e-01
.float  9.28506080e-01
.float  9.32992799e-01
.float  9.37339012e-01
.float  9.41544065e-01
.float  9.45607325e-01
.float  9.49528181e-01
.float  9.53306040e-01
.float  9.56940336e-01
.float  9.60430519e-01
.float  9.63776066e-01
.float  9.66976471e-01
.float  9.70031253e-01
.float  9.72939952e-01
.float  9.75702130e-01
.float  9.78317371e-01
.float  9.80785280e-01
.float  9.83105487e-01
.float  9.85277642e-01
.float  9.87301418e-01
.float  9.89176510e-01
.float  9.90902635e-01
.float  9.92479535e-01
.float  9.93906970e-01
.float  9.95184727e-01
.float  9.96312612e-01
```

```
        .float  9.97290457e-01
        .float  9.98118113e-01
        .float  9.98795456e-01
        .float  9.99322385e-01
        .float  9.99698819e-01
        .float  9.99924702e-01
```

D.2 MOVE.ASM

```
***********************************************************************
*                                                                     *
*                     DMA transfer Program - MOVE.ASM                 *
*                     -------------------------------                 *
*                                                                     *
* host interface program that moves data from evmboard to host.       *
* Using the utility is as easy as step 1 2 3:                         *
* In your program:                                                    *
*       STEP 1) insert line:                                          *
*               .global start_dma                                     *
*       STEP 2) when the data are ready, in your main program add:    *
*               ldi start address of the data block to be moved, R6   *
*               ldi length of the data block, R7                      *
*               call    start_dma                                     *
* In your linker command file:                                        *
*       STEP 3) add line:                                             *
*               move.obj                                              *
* Limitations:                                                        *
* start_dma only does the necessary setup for the data transfer, so   *
* you have to allow sufficient time for the transfer to complete      *
* even after start_dma returns.                                       *
* You have to make sure your linker command file puts .text section   *
* within the reach of DP. And don't let your program or data take up  *
* the dma interrupt vector location ( 0bh ).                          *
* The program can only transfer the lower 16 bits of each data        *
* location.                                                           *
*                                                                     *
* Features ( or Bugs? :-) ):                                          *
* start_dma will do nothing and return if either the host is not      *
* ready or the previous dma session is not complete.                  *
*                                                                     *
*      J Chen and H. Sorensen.                                        *
*      This  code is mostly written by TI                             *
*      Written Aug. 12, 1992. Modified Feb 10, 1995. Version 1.0      *
*                                                                     *
```

```
*  Do not distribute without permission from authors. Copyright 1995 *
*                                                                    *
**********************************************************************
          .global start_dma

V_ADDR     .set    11
MASK       .set    5

           .text
enbl_eint2 .word   000040400h            ;int2 dma (host writes)
dma_rctl   .word   0C0000A13h            ;dma read control
                                         ;c30 mem -> com reg
                                         ;interrupt driven from host
dma_ctl    .word   000808000h            ;dma global control register
hostport   .word   000804000h            ;host interface port address
dma_vector .word   dmadone

start_dma: push    r0
           pushf   r0
           push    ar0

           tstb    1h,if                 ;check if host is ready
           bz      done                  ;to receive
           ldi     @dma_ctl,ar0
           ldi     *+ar0(8),r0
           bnz     done
           ldi     @dma_vector,r0        ;set dma interrupt vector
           ldi     V_ADDR,ar0
           sti     r0,*ar0
           ldi     MASK,r0
           andn    r0,if
           ldi     @hostport,ar0
           sti     r7,*ar0               ;send length of data block
           ldi     @dma_ctl,ar0          ;get dma control address
           sti     r6,*+ar0(4)           ;set dma source address
           ldi     @hostport,r0          ;load host port address
           sti     r0,*+ar0(6)           ;store dma destination address
           sti     r7,*+ar0(8)           ;store dma count value
           ldi     @dma_rctl,r0          ;fetch dma control word
           sti     r0,*ar0               ;store dma control register
           or      @enbl_eint2,ie        ;enable dma read interrupt
done:      pop     ar0
           popf    r0
           pop     r0
           rets
```

```
dmadone:    push    st                  ;save registers
            push    r0
            pushf   r0
            push    ar0

            ldi     @dma_ctl,ar0        ;get dma control address
            xor     r0,r0
            sti     r0,*ar0             ;turn off dma

            pop     ar0
            popf    r0
            pop     r0
            pop     st
            reti
            .end
```

D.3 FFTDISP.C

```
/************************************************************************/
/*                                                                      */
/*              Display frequency response FFT_DISP.C                   */
/*              -------------------------------------                   */
/*                                                                      */
/*  Program to display frequency response on PC screen.                 */
/*  For this program to work, the EVM input should be a sinusoidal      */
/*  input sweeping from 50 Hz to 4 kHz.                                 */
/*                                                                      */
/*      Jianping Chen,  Henrik Sorensen.                                */
/*      Written in Aug. 1992                              Version 1.0 */
/*                                                                      */
/*  Do not distribute without permission from authors.                  */
/*                                                      Copyright 1993-96 */
/*                                                                      */
/************************************************************************/
#include <stdio.h>       /* include necessary standard header files     */
#include <stdlib.h>
#include <ctype.h>
#include <time.h>
#include <graphics.h>    /* include TURBO C specific header files       */
#include <conio.h>
#include <bios.h>
#include <math.h>
#include "pc_1.h"
```

```
#define  PI 3.14159265
#define  Fsamp 8000.0
#define  HEIGHT 300.

int     GraphDriver;
int     GraphMode;
int     data_buffer[512],length,y[512];

void  init_graphics(void);
void  display_error(void);
void  get_buffer(void);
void  graph(void);
void  axis(void);

main()
{
    int i;
    char ch;

    init_evm();          /* initialize EVM                                   */
    init_graphics();
    setbkcolor(1);
    axis();
    setviewport( 64, 25, 575, 324, ON );
    setfillstyle(SOLID_FILL,3);
    bar( 0,0,511,299 );
    for ( i=0; i<512; i++ ) y[i] = HEIGHT;
    do
    {
        start_command();
        get_buffer();
        graph();
        if ( kbhit() )
        {
                ch = bioskey(0);
                if ( ch == 'q' || ch == 'Q' )
                        break;
        }
    }
    while( 1 );
    textmode( LASTMODE );
    clrscr();
    return(1);
}
```

```
void init_graphics(void)
{
    char ch ;

    GraphDriver = GraphMode = DETECT;
                                      /* what graphics driver present? */
    detectgraph(&GraphDriver, &GraphMode);
    if ((GraphDriver == EGA) || (GraphDriver == VGA))
    {
         GraphMode = EGAHI;
         registerbgidriver(EGAVGA_driver);
    }
    else if (GraphDriver == CGA)
        {
            printf("Is this a Compaq w/Plasma Display? (Y or N):");
            ch = getche();
            if ( (ch == 'y') || (ch == 'Y') )
            {
               GraphDriver = ATT400; GraphMode = ATT400HI;
               registerbgidriver(ATT_driver);
            }
            else
            {
               display_error();
            }
        }
        else
            display_error();

    initgraph(&GraphDriver, &GraphMode, "");
    if ( graphresult() != grOk )   display_error();

/*********************************************************************/
    /*---------------------------------------------------------------*/
    /* MAKE SURE DISPLAY HAS ADEQUATE RESOLUTION                     */
    /*---------------------------------------------------------------*/
    if (getmaxx() < 639 || getmaxy() < 349)  display_error();

    /*---------------------------------------------------------------*/
    /* DRAW ALL PERMANANT / NON-CHANGING WINDOWS                     */
    /*---------------------------------------------------------------*/

    setcolor(WHITE);           /* set text and border color to white */
```

```
    if (GraphDriver == ATT400)
        setfillstyle(SOLID_FILL,BLACK); /* set window fill color     */
    else
        setfillstyle(SOLID_FILL,BLUE);
}

void display_error(void)
{
    printf("\nNeed VGA, EGA, or AT&T 640x400 Display\n");
    exit( 1 );
}

void get_buffer(void)
{
    int i;

    length = receive_data();
    for ( i=0; i<length; i++ )
    {
        data_buffer[i] = receive_data();
    }
}

void graph(void)
{
    int idx, tmp;

    for ( idx=0; idx<length; idx++ )
    {
       tmp = y[idx];
       y[idx] = HEIGHT - data_buffer[idx] * HEIGHT / 4096.;
       setcolor( (y[idx] > tmp)? 3 : 4 );
       line(512/length*idx, y[idx], 512/length*idx, tmp );
    }
}

void axis(void)
{
    int i;
    char    *xlabel[9]=
    {
        "0", "1K", "2K", "3K", "4K(Fs/2)", "", "", "", "Fs"
    };

    setviewport( 64, 325, 585, 345, ON );
```

```
    setcolor(YELLOW);
    line(0, 0, 512, 0);
    setcolor(WHITE);
    for ( i=0; i<=512; i+=32 )
    {
         line(i, 0, i, 4);
    }
    outtextxy( 0, 13, xlabel[0] );
    for ( i=0; i<=512; i+=64 )
    {
        line(i, 0, i, 8);
        outtextxy( i-8, 13, xlabel[i/64] );
    }
}
```

Bibliography

[1] CCITT. *Blue Book, Vol. VIII, Fas. VIII.1: Data Communication on the Telephone Network.* ITU, Geneva, 1988.

[2] R. Chassaing and D.W. Horning. *Digital Signal Processing with the TMS320C25.* John Wiley, New York, NY, 1990.

[3] R.E. Crochiere and L.R. Rabiner. *Multirate Digital Signal Processing.* Prentice Hall. Inc., Englewood Cliffs, NJ, 1983.

[4] Pickholtz et al, "Theory of spread spectrum communication: A tutorial," *IEEE Transactions on Communication*, vol. COM-30, no. 5, pp. 855–884, May 1982.

[5] B.A. Hutchins and T.W. Parks. *A Digital Signal Processing Laboratory using the TMS320C25.* Prentice-Hall, Englewood Cliffs, NJ, 1990.

[6] D.L. Jones and T.W. Parks. *A Digital Signal Processing Laboratory using the TMS32010.* Prentice-Hall, Englewood Cliffs, NJ, 1988.

[7] MathWorks, Inc. *Pro-MatLab, User's Guide.* The MathWorks, Inc., 24 Prime Park Way, Natick, MA 01760, 1991.

[8] S.K. Mitra and J.F. Kaiser. *Handbook for Digital Signal Processing.* John Wiley & Sons, New York, NY, 1993.

[9] A.V. Oppenheim and R.W. Schafer. *Discrete-time Signal Processing.* Prentice-Hall, Englewood Cliffs, NJ, 1989.

[10] A.V. Oppenheim and C. Weinstein, "A bound on the output of a circular convolution with application to digital filtering," *IEEE Trans. Audio Electroacoust.*, vol. AU-17, no. 2, pp. 120–124, Jun. 1969. Reprinted in *Digital Signal Processing*, ed. L. R. Rabiner and C. M. Rader, pp. 344-348, New York: IEEE Press, 1972.

[11] T.W. Parks and C.S. Burrus. *Digital Filter Design.* John Wiley & Sons, New York, NY, 1987.

[12] John G. Proakis and Masoud Salehi. *Communication Systems Engineering.* Prentice Hall, Englewood Cliffs, NJ, 1994.

[13] H.V. Sorensen, D.L. Jones, M.T. Heideman, and C.S. Burrus, "Real-valued fast Fourier transform algorithms," *IEEE Trans. Acoust., Speech, and Signal Processing*, vol. ASSP-35, no. 6, pp. 849–863, June 1987.

[14] Texas Instruments. *Digital Signal Processing Applications with the TMS320C30 Evaluation Module.* Texas Instruments, Dallas, TX, 1990.

[15] Texas Instruments. *TMS320C30 Evaluation Module Technical Reference.* Texas Instruments, Dallas, TX, 1990.

[16] Texas Instruments. *TMS320C30 Optimizing C Compiler Reference Guide.* Texas Instruments, Dallas, TX, 1990.

[17] Texas Instruments. *TMS320C30 Assembly Language Tools User's Guide.* Texas Instruments, Dallas, TX, 1991.

[18] Texas Instruments. *TMS320C3x C Source Debugger User's Guide.* Texas Instruments, Dallas, TX, 1991.

[19] Texas Instruments. *TMS320C3x User's Guide.* Texas Instruments, Dallas, TX, 1991.

[20] P.P. Vaidyanathan, "Quadrature mirror filter banks, m-band extensions and perfect-reconstruction techniques," *IEEE ASSP Magazine*, vol. 4, no. 3, pp. 4–20, Jul. 1987.

[21] B. Widrow and S.D. Stearns. *Adaptive Signal Processing.* Prentice Hall. Inc., Englewood Cliffs, NJ, 1985.

[22] R.E. Ziemer and R.L. Peterson. *Digital Communication and Spread Spectrum Systems.* McMillan and Collier, New York, NY, 1985.

Index

YOU SHOULD CAREFULLY READ THE FOLLOWING TERMS AND CONDITIONS BEFORE OPENING THIS DISKETTE PACKAGE. OPENING THIS DISKETTE PACKAGE INDICATES YOUR ACCEPTANCE OF THESE TERMS AND CONDITIONS. IF YOU DO NOT AGREE WITH THEM, YOU SHOULD PROMPTLY RETURN THE PACKAGE UNOPENED, AND YOUR MONEY WILL BE REFUNDED.

IT IS A VIOLATION OF COPYRIGHT LAWS TO MAKE A COPY OF THE ACCOMPANYING SOFTWARE EXCEPT FOR BACKUP PURPOSES TO GUARD AGAINST ACCIDENTAL LOSS OR DAMAGE

Prentice-Hall, Inc. provides this program and licenses its use. You assume responsibility for the selection of the program to achieve your intended results, and for the installation, use, and results obtained from the program. This license extends only to use of the program in the United States or countries in which the program is marketed by duly authorized distributors.

LICENSE

You may:

a. use the program;
b. copy the program into any machine-readable form without limit;
c. modify the program and/or merge it into another program in support of your use of the program.

LIMITED WARRANTY

THE PROGRAM IS PROVIDED "AS IS" WITHOUT WARRANTY OF ANY KIND, EITHER EXPRESSED OR IMPLIED, INCLUDING, BUT NOT LIMITED TO, THE IMPLIED WARRANTIES OF MERCHANTABILITY AND FITNESS FOR A PARTICULAR PURPOSE. THE ENTIRE RISK AS TO THE QUALITY AND PERFORMANCE OF THE PROGRAM IS WITH YOU. SHOULD THE PROGRAM PROVE DEFECTIVE, YOU (AND NOT PRENTICE-HALL, INC. OR ANY AUTHORIZED DISTRIBUTOR) ASSUME THE ENTIRE COST OF ALL NECESSARY SERVICING, REPAIR, OR CORRECTION.

SOME STATES DO NOT ALLOW THE EXCLUSION OF IMPLIED WARRANTIES, SO THE ABOVE EXCLUSION MAY NOT APPLY TO YOU. THIS WARRANTY GIVES YOU SPECIFIC LEGAL RIGHTS AND YOU MAY ALSO
HAVE OTHER RIGHTS THAT VARY FROM STATE TO STATE.

Prentice-Hall, Inc. does not warrant that the functions contained in the program will meet your requirements or that the operation of the program will be uninterrupted or error free.

However, Prentice-Hall, Inc., warrants the diskette(s) on which the program is furnished to be free from defects in materials and workmanship under normal use for a period of ninety (90) days from the date of delivery to you as evidenced by a copy of your receipt.

LIMITATIONS OF REMEDIES

Prentice-Hall's entire liability and your exclusive remedy shall be:

1. the replacement of any diskette not meeting Prentice-Hall's "Limited Warranty" and that is returned to Prentice-Hall with a copy of your purchase order, or

2. if Prentice-Hall is unable to deliver a replacement diskette or cassette that is free of defects in materials or workmanship, you may terminate this Agreement by returning the program, and your money will be refunded.

IN NO EVENT WILL PRENTICE-HALL BE LIABLE TO YOU FOR ANY DAMAGES, INCLUDING ANY LOST PROFITS, LOST SAVINGS, OR OTHER INCIDENTAL OR CONSEQUENTIAL DAMAGES ARISING OUT OF THE USE OR INABILITY TO USE SUCH PROGRAM EVEN IF PRENTICE-HALL, OR AN AUTHORIZED DISTRIBUTOR HAS BEEN ADVISED OF THE POSSIBILITY OF SUCH DAMAGES, OR FOR ANY CLAIM BY ANY OTHER PARTY.

SOME STATES DO NOT ALLOW THE LIMITATION OR EXCLUSION OF LIABILITY FOR INCIDENTAL OR CONSEQUENTIAL DAMAGES, SO THE ABOVE LIMITATION OR EXCLUSION MAY NOT APPLY TO YOU.

GENERAL

You may not sublicense, assign, or transfer the license or the program except as expressly provided in this Agreement. Any attempt otherwise to sublicense, assign, or transfer any of the rights, duties, or obligations hereunder is void.

This Agreement will be governed by the laws of the State of New York.

Should you have any questions concerning this Agreement, you may contact Prentice-Hall, Inc., by writing to:

Prentice Hall
Engineering and Computer Science
College Division
One Lake Street
Upper Saddle River, New Jersey 07458

YOU ACKNOWLEDGE THAT YOU HAVE READ THIS AGREEMENT, UNDERSTAND IT, AND AGREE TO BE BOUND BY ITS TERMS AND CONDITIONS. YOU FURTHER AGREE THAT IT IS THE COMPLETE AND EXCLUSIVE STATEMENT OF THE AGREEMENT BETWEEN US THAT SUPERSEDES ANY PROPOSAL OR PRIOR AGREEMENT, ORAL OR WRITTEN, AND ANY OTHER COMMUNICATIONS BETWEEN US RELATING TO THE SUBJECT MATTER OF THIS AGREEMENT.

ISBN 0-13-741828-0